Discrete Model for Pattern Formation
in Laser-Induced Jet-Chemical Etching

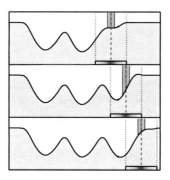

Von der Fakultät Mathematik und Physik der Universität Stuttgart
zur Erlangung der Würde eines Doktors der
Naturwissenschaften (Dr. rer. nat.) genehmigte Abhandlung

Vorgelegt von

M. Sc. Alejandro Mora

aus Bogotá (Kolumbien)

Hauptberichter: Prof. Dr. rer. nat. H.J. Herrmann
Mitberichter: Prof. Dr. rer. nat. H.-R. Trebin
Tag der mündlichen Prüfung: 30. Januar 2006

Institut für Höchsleistungsrechnen (IHR) der Universität Stuttgart
2006

Bibliografische Information Der Deutschen Bibliothek

Die Deutsche Bibliothek verzeichnet diese Publikation in der Deutschen
Nationalbibliografie; detaillierte bibliografische Daten sind im Internet über
http://dnb.ddb.de abrufbar.

ISBN 3-8325-1193-8

Logos Verlag Berlin
Comeniushof, Gubener Str. 47,
10243 Berlin
Tel.: +49 030 42 85 10 90
Fax: +49 030 42 85 10 92
INTERNET: http://www.logos-verlag.de

To my mother and my sisters for their love.

To Andrea, who makes every single day
the happiest day of my life.

Acknowledgements

First at all I am deeply thankful to Dr. Maria Haase for giving me this generous and unique opportunity. Without her guidance, support, and perseverance this work would not exist. I wish to express my gratitude to Prof. Dr. Hans Herrmann for the his advice and support. I am thankful to Prof. Dr. Hans-Rainer Trebin who kindly agreed to be examiner of this thesis.

This work was part of the project " *Non-linear dynamics of the laser based electrochemical jet method for the ultra-precision micro-structuring of metals for applications in optics*" supported by the Volkswagen Foundation. For providing such stimulating research cooperation, I am very grateful to Prof. Dr. P. Plath and T. Rabbow (IAPC - Bremen University); Prof. Dr. S. Metev and A. Stephen (BIAS - Bremen); Prof. Dr. J. Peinke and J. Linke (Hydro - Oldenburg University).

My sincere gratitude for all the help, support, and encouragement to all the people of the IHR (former ICA-CSV) institute: M. Haase, P. Streiner, M. Dziobek and P. Hermann. I thank Prof. Dr. M. Resch (IHR) for the support during the last stage of this work. I am indebted with the collegues of the ICP (former ICA1) institute for the encouragement and scientific input.

This work would not be possible without the influence of ideas, work and wisdom of many people. I am specially thankful with T. Rabbow for the experimental data, the discussions, and his nice encouragement. I thank A. Stephen for the discussions and some of the figures of Chapter 2. I gratefully acknowledge Prof. Dr. R. Friedrich for his valuable contributions. I thank C. Gerlach, M. Wächter, J. Peinke, G. Radons, S. Linz, A. Kouzmitchev, B. Lehle, F. Kun, J. D. Muñoz, F. Reischel, P. Lind, J. Gallas, and S. Eberth for interesting discussions and contributions.

I thank my mother and sisters who accompanied me in this enterprise. I thank Andrea Ehl for her love and daily encouragement and motivation. I would like to say a sincere "muchas gracias" to all good friends, they made these years unforgettable: F. Fonseca, F. Alonso, A. Ehl, L. Rosero, M. C. González, D. Alonso, R. Cruz, V. Schwaemmle, E. Partelli, M. Sidri, J. Garibay, L. López, O. Pozos, L. Aquilera, J. Jovel, N. Bludau, C. Pereira, M. Cuervo, L. Zenteno, U. Habel, M. C. Blanco, A. Jeldrez, O. Durán, P. Lind, M. Lingenfelder, D. Repetto, C. Russo, R. Biondi, J. Rodriguez, A. Peña, and H. Mendez. Thanks for all the fun to the people of the OLA organization, Bail-OLA group, and "Colombia Candela" group.

I am deeply grateful to people and government of Germany for giving me the opportunity to know and enjoy this great country. This work was funded by the Volkswagen-Stiftung grant I/77315.

Alejandro Mora

Stuttgart, February 2006

Deutsche Zusammenfassung

Während der letzten Jahre ist die Ultrapräzisionsmikrostrukturierung von Metallen wichtiger Gegenstand intensiver Forschungen gewesen. Zu den stetig wachsenden, zukunftsweisenden Anwendungsfeldern gehören unter anderem die Produktion von metallischen Werkzeugeinsätzen für die Mikroabformtechnik und die Herstellung optischer Komponenten, von Instrumenten für die minimal-invasive Mikrochirurgie und für Anwendungen der Strömungstechnik in der Biomedizin sowie von mikroelektronisch-mechanischen Systemen. Wichtigster Aspekt vom experimentellen und theoretischen Standpunkt aus ist dabei die Kontrolle der Qualität und der Topographie der mikrostrukturierten Oberflächen. Um unerwünschte Effekte durch lokal hohe Laserintensitäten beim Laserschmelzschneiden zu vermeiden, wurden verschiedene laserverstärkte elektrochemische Abtragstechniken vorgeschlagen [Datta, 1998a,b]. In der vorliegenden Arbeit untersuchen wir die Oberflächenstrukturen, die bei Anwendung eines vor kurzem entwickelten lasergestützten chemischen Jet-Verfahrens zum Ätzen von Metallen entstehen (laser- induced jet- chemical etching (LJE)) [Metev *et al.*, 2003; Nowak & Metev, 1996; Rabbow *et al.*, 2005; Stephen *et al.*, 2002, 2004]. In diesem Versuchsaufbau werden Stahlfolien mit Hilfe eines fokussierten Laserstrahls in Kombination mit einem Ätzmitteljet auf Mikroskala strukturiert.

Bei Zimmertemperatur wird die metallische Oberfläche, die mit dem Ätzmittel in Berührung kommt, in sehr kurzer Zeit passiviert (passivated) und damit von weiteren Ätzreaktionen abgehalten. Durch Absorption der Laserstrahlung und Wärmeausbreitung wird ein Bereich der Oberfläche erhitzt, der weitaus größer ist als der Laserspot. Oberhalb einer gewissen Grenztemperatur wird die Passivierungsschicht aufgebrochen und es beginnt thermisch aktiviertes Ätzen. Die Aufgabe des Ätzmitteljets ist es, frisches Ätzmittel zuzuführen und den Transport von gelöstem Material zu beschleunigen. Da das metallische Werkstück relativ zum Laser mit einer bestimmten Vorschubgeschwindigkeit bewegt wird, entsteht durch die Ätzfront ein Graben auf der Oberfläche. Das Oberflächenprofil dieses Grabens hängt von der Vorschubgeschwindigkeit, der Laserleistung, der Konzentration des Ätzmittels und der Geschwindigkeit des Ätzmitteljets ab.

In den meisten Fällen bildet sich eine stationäre Ätzfront aus und es entstehen Gräben von nahezu konstanter Breite und Tiefe, die durch den Ätzvorgang eine gewisse Oberflächenrauigkeit aufweisen. Auf der anderen Seite entstehen für gewisse Bereiche der Vorschubgeschwindigkeit, der Laserleistung und der Geschwindigkeit des Ätzmitteljets Oszillationen in der Ätzfront, die periodische Rippelstrukturen zur Folge haben [Rabbow et al., 2005]. In dieser Arbeit betrachten wir den Fall, dass die Rippels intrinsisch generiert werden und zu einer universellen Klasse von Musterbildungsprozessen gehören, die beispielsweise beobachtet werden, wenn Wind über Sand streicht, beim Ionenbeschuss verschiedener Oberflächen [Makeev et al., 2002], [Valbusa et al., 2002] und beim Wasserstrahlschneiden [Friedrich et al., 2000b]. Kapitel 3 befasst sich mit der Morphologie einer Reihe von Gräben, die bei verschiedenen Vorschubgeschwindigkeiten strukturiert wurden. Mit Hilfe klassischer Fouriertechniken waren wir in der Lage, zwischen intrinsisch generierten und durch äußere Oszillationen induzierten Rippelstrukturen zu unterscheiden.

Will man die Prozesse modellieren, die zu der kleinskaligen Rauigkeit der Oberflächen führen und mit experimentellen Messungen vergleichen so werden dazu numerische Methoden benötigt, die eine möglichst vollständige quantitative Charakterisierung der Oberflächen ermöglichen. In Kapitel 4 fassen wir unsere früheren Arbeiten hinsichtlich der Charakterisierung der Oberflächenrauigkeit von Werkstücke zusammen [Haase et al., 2004; Mora & Haase, 2004a,b; Mora et al., 2004]. Darüber hinaus präsentieren wir einige Ergebnisse, die wir durch Anwendung einer neuen, auf der Markov-Analyse basierenden stochastischen Methode zur vollständigen Beschreibung der Skalenabhängigkeit der Rauigkeit von elektropolierten Oberflächen erzielt haben. Letztendlich könnte man mit diesem Verfahren auch die Strukturen auf dem Boden und an den Wänden der Gräben analysieren. Dazu wären jedoch wesentlich mehr Messdaten erforderlich.

Es ist ein große Herausforderung, ein mikroskopisches Modell zu erstellen, das der Dynamik aller beteiligten Prozesse auf den verschiedenen räumlichen und zeitlichen Skalen gerecht wird. Die nichtlineare Dynamik der Absorption des Laserstrahls, der Wärmeausbreitung, der chemischen Reaktionen, der Hydrodynamik und der Transportphänomene macht das System zu komplex, um es vollständig zu modellieren. Durch die permanente Einwirkung des Laserstrahls und die Zuführung des Ätzmittels muss das LJE-Experiment als ein System fern vom thermodynamischen Gleichgewicht behandelt werden. Zur Untersuchung der Musterbildung in solchen komplexen Systemen greift man gewöhnlich auf phänomenologische Modelle bzw. Modellgleichungen zurück. Eine dieser Gleichungen ist die Kuramoto-Sivashinsky(KS) Gleichung [Kuramoto, 1984; Kuramoto & Tsuzuki, 1976; Sivashinsky, 1979]. Diese Gleichung wurde als allgemeine nichtlineare Kontinuumsbeschreibung für verschiedene Musterbildungsprozesse vorgeschlagen. Ein typisches Merkmal dieses Systems ist, dass sich im Anfangsstadium periodische Strukturen ausbilden, die jedoch nach einiger Zeit zerfallen und einem ungeordneten Zustand Platz machen. Wir werden zeigen, dass die Musterbildung beim Ionenbeschuss, bei Wasserstrahlschneiden und beim LJE-Experiment universellen Charakter hat und mit stochastischen Differentialgleichungen beschrieben werden kann, die eng mit der Kuramoto-Sivashinsky-Gleichung zusammenhängen (Kapitel 6 und 7).

Anstatt ein mikroskopisches Modell unter Berücksichtigung des Zusammenspiels aller chemischen und physikalischen Prozesse zu formulieren, schlagen wir ein minimales phänomenologisches Modell zur Beschreibung der Dynamik der geätzten Grenzfläche vor, das auf Ideen beruht, die aus der Theorie der Evolution von Oberflächen in Systemen fern vom thermodynamischen Gleichgewicht stammen. In der Literatur wurden eine ganze Reihe von diskreten Wachstumsmodellen und stochastischen Differentialgleichungen vorgeschlagen, die das kinetische Aufrauen beim Wachstum oder der Erosion von Oberflächen beschreiben [Barabási & Stanley, 1995].

In dieser Arbeit wird die Simulation der Musterbildungsphänome (pattern formation phenomena), die beim LJE-Experiment auftreten in zwei Schritten ausgeführt. Zunächst wird ein *erweitertes Modell* (extended model) entwickelt und im Anschluss daran wird die eigentliche *LJE-Simulation* durchgeführt (Kapitel 7 und 8). Das erweiterte Modell ist eine Modifikation eines diskreten Modells von Cuerno, Makse, Tommassone, Harrington, and Stanley (CMTHS) [Cuerno *et al.*, 1995]. Beim CMTHS-Modell wird die Evolution einer ebenen Oberfläche simuliert, die gleichmäßig mit einem Ionenstrahl beschossen wird. Ausgehend von der Annahme von zugrunde liegenden universellen Gesetzmäßigkeiten stellen wir ein erweitertes Modell vor, um das instabile Verhalten eines gleichförmigen Ätzens einer ebenen metallischen Oberfläche zu simulieren. Beim LJE-Experiment handelt es sich jedoch um eine thermisch aktivierte, lokalisierte Mikrostrukturtechnik. Deshalb beruht die LJE-Simulation der Grabenstruktur auf einer Kombination des erweiterten Modells mit der Approximation thermisch dünner Schichten (thermally thin layer approximation), um die Temperaturabhängigkeit und das inhomogene Ätzen der LJE-Technik zu berücksichtigen.

Die Dynamik des CMTHS und des erweiterten Modells wird in Kapitel 7 charakterisiert. Die Oberfläche wird durch ein diskretes Gitter repräsentiert, dessen Evolution durch das Wechselspiel zwischen Erosion und Diffusionsprozessen an zufällig ausgewählten Stellen des Gitters angetrieben wird. Bei der Erosion werden in Abhängigkeit von der Krümmung der Oberfläche einzelne Zellen des Gitters entfernt. Die Oberflächendiffusionsprozesse werden durch Sprünge auf benachbarte Gitterstellen repräsentiert. Im Rahmen der kinetischen Monte-Carlo-Methode werden solche zufälligen Ereignisse mit einer bestimmten Ereigniswahrscheinlichkeit simuliert [Newman & Barkema, 1999]. Zur Abschätzung dieser Wahrscheinlichkeitsraten werden alle zur Verfügung stehenden Informationen herangezogen, die man aus experimentellen Beobachtungen und theoretischen Überlegungen gewinnen kann [Metiu et al., 1992].

Beim CMTHS-Modell ist die Evolution einer ursprünglich ebenen Oberfläche charakterisiert durch ein anfängliches Aufrauen, auf das ein kurzes Regime folgt, in dem sich Rippelstrukturen ausbilden. Infolge nichtlinearer Einflüsse werden diese periodischen Strukturen jedoch bald zu einem späteren Zeitpunkt wieder zerstört und es entsteht erneut eine raue Topographie. Das CMTHS-Modell hat nur begrenzte Möglichkeiten Rippels auszubilden. Daher benutzen wir ein erweitertes Modell, das auf einer genaueren Abschätzung der lokalen Krümmung für die Erosionsregel beruht [Mora et al., 2005a,b]. Die Entwicklung der Oberfläche im erweiterten Modell verläuft ähnlich wie beim CMTHS-Modell, wobei allerdings das Rippel-Regime über längere Zeiträume erhalten bleibt. Außerdem sind die Rippel-Strukturen regulärer und ihre Wellenlänge ist größer als beim CMTHS-Modell; eine notwendige Voraussetzung für die Simulation von lokalem Ätzen. Beim erweiterten Modell werden die Rippelstrukturen im Lauf der Zeit gröber, d.h. die Wellenlänge nimmt im Lauf der Evolution zu (coarsening).

Um die analytischen Eigenschaften des erweiterten Modell besser zu verstehen, haben wir den Zusammenhang mit einer kontinuierlichen Beschreibung durch eine stochastische Differentialgleichung untersucht. In Anlehnung an eine Methode, die von Vvedensky und Mitarbeitern vorgeschlagen wurde (siehe [Chua et al., 2005] und die darin zitierten Referenzen) und die auf der Master-Gleichung basiert zeigen wir, dass dem erweiterten Modell im Kontiuum-Limit eine modifizierte Kuramoto-Sivashinsky (KS) Gleichung entspricht. Diese Gleichung entsteht durch eine Gradienten-Entwicklung, wobei die Terme niedriger Ordnung der KS-Gleichung entsprechen, während höhere Gradienten für das Phänomen der Vergröberung der Rippelstrukturen verantwortlich zu machen sind. Durch die Inhomogenität der Ätzfront, die durch die Bewegung des Laserkopfs relativ zur Metalloberfläche entsteht, ergibt sich in einem mit dem Laser mitbewegten Koordinatensystem eine modifizierte KS-Gleichung [Mora *et al.*, 2005b].

Die Strukturbildung beim LJE-Prozess wird in erster Linie durch die Erwärmung der Oberfläche infolge der Laserabsorption bedingt. Die von uns vorgeschlagene Approximation des Temperaturfelds für dünne Schichten (thermally thin layer) liefert eine sinnvolle Abhängigkeit des Temperaturfelds von der Laserleistung. Aufgrund dieses Temperaturfelds und unter Berücksichtigung der Tatsache, dass der Ätzprozess erst oberhalb einer gewissen Temperaturschwelle beginnt, definieren wir in Kapitel 8 eine Wahrscheinlichkeitsverteilung für das Ätzen (etching probability distribution), das die räumliche Abhängigkeit der Erosionsraten für das erweiterte Modell bestimmt. Mit Hilfe dieser Wahrscheinlichkeitsverteilung ist es möglich, die Tatsache zu berücksichtigen, dass nur lokal geätzt wird, dass die Abtragsrate von der Temperatur abhängt und dass die Ätzfront wesentlich breiter ist als der Laserspot.

Bei der LJE-Simulation wird die Wahrscheinlichkeitsverteilung für das Ätzen mit einer bestimmten Vorschubgeschwindigkeit relativ zum Gitter bewegt und für die Abtragsraten bei Anwendung der Erosionsregeln verwendet. Daher ist bei der Entstehung eines Grabens jeder Teil der Oberfläche nur so lange dem Strukturierungsprozess ausgesetzt, wie der Laser zum Überstreichen benötigt. Diese Verweildauer hängt von der Vorschubgeschwindigkeit und dem Durchmesser der Wahrscheinlichkeitsverteilung des Ätzprozesses ab. Ist die Vorschubgeschwindigkeit groß, so ist die Verweildauer kurz und der Boden des Grabens zeigt eine Topographie, die der Rauigkeit des erweiterten Modells in einem frühen Stadium entspricht. Reduziert man die Vorschubgeschwindigkeit, so kann man (in bestimmten Parameterbereichen) eine Topographie beobachten, die einem Rippleregime entspricht. Ist die Vorschubgeschwindigkeit jedoch zu klein, so entsteht durch die lange Verweildauer ein Graben mit einer rauen Topographie am Boden, die derjenigen in einem späten Stadium des erweiterten Modells entspricht. Zusammenfassend werden in der LJE-Simulation in Abhängigkeit von der Verweildauer Gräben verschiedener Topographie strukturiert, die den unterschiedlichen Stadien der Entwicklung des erweiterten Modells entsprechen.

Die LJE-Simulation ermöglicht es daher, für verschiedene Bereiche der Vorschubgeschwindigkeit und der Laserleistung gerippelte Gräben zu erzeugen. Die Abhängigkeit der Rippellänge von der Vorschubgeschwindigkeit und der Laserleistung stimmt qualitativ mit den experimentellen Beobachtungen überein. Das hierfür verwendete diskrete statistische Modell ermöglicht es auf einfache Weise, die grundlegenden physikalischen und chemischen Prozesse und wesentlichen Parameter die bei der LJE-Technik auftreten, abzubilden. Die Simulationen geben Einblick in die inneren Mechanismen, die zur Rippelbildung führen, und liefern Anregungen für weitere Experimente. Letztendlich ist es da Ziel, ein leicht verständliches Modell zu erstellen, das die physikalischen und chemischen Prozesse wiedergibt, die sich an der Oberfläche abspielen, wobei auch der Einfluss des Ätzmittel-Jets und des gesamten Materials der Metallfolie berücksichtigt wird. Ein grundlegendes Verständnis des Systems kann es daher bei zukünftigen technischen Anwendungen ermöglichen, die gewünschten Oberflächentopographien gezielt anzusteuern. Weiterentwicklungen und Modifikationen des LJE-Modells könnten ferner dazu verwendet werden, die Musterbildung bei anderen Schneidetechniken wie dem Wasserstrahlschneiden oder dem Laserschmelzschneiden zu simulieren.

Insgesamt wurde in dieser Dissertation gezeigt, wie man mit Hilfe diskreter stochastischer Modelle die nichtlineare Dynamik und Musterbildung in Systemen fern vom thermodynamischen Gleichgewicht simulieren kann. Obwohl man die Details der dabei beteiligten wechselwirkenden Prozesse nicht kennt, können (aufgrund der Universalität der Musterbildung) doch die grundlegenden physikalischen und chemischen Prozesse durch einfache Mechanismen modelliert werden und geben so Einblick in die komplexe Dynamik des Systems. Diese Klasse von phänomenologischen Modellen nimmt eine Mittlerfunktion zwischen Experiment und Theorie ein und kann für Vorhersagen verwendet werden. Weiterentwicklungen der Regeln für das Modell, die die Mechanismen auf mikroskopischer Skala und die Materialeigenschaften realistischer abbilden, könnten zukünftig dafür verwendet werden, um gleichzeitig globale Stukturen auf makroskopischer Skala und Musterbildung sowie Rauigkeit auf mikroskopischer Skala zu erzeugen.

Contents

Chapter 1

Introduction

1.1 The scope of this thesis

This thesis advances on the application of discrete models to the simulation and under-standing of pattern formation occurring in microstructuring techniques. We focus on the issue of ripples appearing in kerfs (grooves) structured on metallic surfaces by *laser-induced jet-chemical etching* (LJE). The system is far from equilibrium and is composed by various chemical and physical processes with non-linear interaction, which is the typ-ical scenario for emergence of patterns.

Our approach is based on the idea that in spite of the complexity of the system, it is pos-sible to envisage few basic mechanisms that determine the main features of the dynamics of the surface. Based on the assumption that those mechanisms are common to different pattern forming systems, we have adapted and extended a model initially proposed for ion beam sputtering for the LJE experiment. Therefore, the complexity emerges by inter-action of simplified representations of the physical processes, namely erosion and surface diffusion rules acting on a discrete lattice. Computer simulations give convincing indi-cations that the model proposed in this work is a successful first step to understand the pattern phenomenon occurring in the LJE experiment.

1

1.2   The LJE technique: complexity and non-linearity

Structuring of materials at increasingly small scales has been one the main quests of scientific and technologic advancement. In particular, the modification of metallic surfaces at micrometer (and submicrometer) scales is considered to be a key future fabrication technology of precise devices and tools. In order to avoid unwanted effects produced by high power laser evaporation techniques, diverse laser-enhanced electrochemical-dissolution techniques have been proposed [Datta, 1998a,b]. In this work we address the study of a recently developed experimental setup called *laser-induced jet-chemical etching* (LJE). [Metev *et al.*, 2003; Nowak & Metev, 1996; Rabbow *et al.*, 2005; Stephen *et al.*, 2002, 2004]). In the experimental setup, foils of stainless steel are structured at micrometer scale by means of the combined action of a focused laser beam and a jet of chemical etchant. Because the metallic sample is moved with respect to laser with a *feed velocity*, an *etching front* is formed structuring a kerf on the surface. Naturally, the resulting topography depends strongly on the feed velocity, laser power, etchant concentration, and velocity of the etchant jet.

In most of the cases the etching front is approximately steady (in the frame moving with the laser beam), and the resulting structure, although rough, is called *uniform kerf*. However, a remarkable behavior appears for certain ranges of power or feed velocity: an unstable etching front creates a kerf with periodic ripples [Rabbow *et al.*, 2005]. The resulting ripple wavelength varies with the power or feed velocity. The formulation of a phenomenological model that exhibits similar behavior is the main objective of this work. One of the main aims is to obtain understanding and control of the pattern formation mechanism in order to prevent the ripples or shaped them for specific applications. We consider that the ripples are of intrinsic nature, which means that they are not a consequence of oscillations produced by external sources. Our analysis is based in the assumption that these ripples belongs to a family of pattern formation phenomena that emerge for example, in ripple structures formed by wind over a sand bed, ion sputtering of various surfaces, abrasive water-jet cutting, and laser ablation techniques.

It is a big challenge to formulate a microscopic model for the LJE technique that describes the dynamics of different processes occurring in a wide range of spatial and temporal scales. The non-linear dynamics of the laser light absorption, heat, chemical reactions, hydrodynamics, and transport phenomena makes the system too complex to be fully modeled. Instead, we propose to simulate the system by the interactions of simple mechanisms that mimic the relevant physical processes.

1.3 Far from equilibrium systems and pattern formation

Due to the permanent influence of the laser beam and jet of etchant, the LJE experiment
is a system far from thermal equilibrium. From the theoretical point of view, in nonequilibrium systems it is not possible to define a "reasonable" energy function (Hamiltonian), and the system is determined by its dynamical rules [Evans, 2000]. This is
in contrast to equilibrium systems where once the Hamiltonian is specified, it is possible to define a Gibbs ensemble from which properties of the system can be computed
[Schmittmann & Zia, 1995].

As spatially extended systems are driven away from equilibrium by external influences,
they tend to display progressively more complicated dynamics, eventually evolving from
spatially uniform states to spatially patterned states. The non-linear interaction among the
small and fast scale constituents of the system give rise to order at larger spatial and longer
temporal scales. This pattern formation phenomena is ubiquitous in many systems, for
example, coupled reaction-diffusion systems, vertically vibrated fluids, Rayleigh-Bérnard
convection, patterns in non-linear optics, etc. (see reviews in [Cross & Hohenberg, 1993;
Walgraef, 1997]). They have been studied extensively in the framework of theories of
dissipative structures [Nicolis & Prigogine, 1977], synergetics [Haken, 1977], and self-
organization [Krinsky, 1984].

It is usual to find that the formulation of "microscopic" partial differential equations
(PDE) for many of those pattern forming systems is not possible (microscopic in the
sense of elementary building blocks)[Cross & Hohenberg, 1993]. Moreover, if the exact equations, boundary conditions, and corresponding parameters can be determined
for a specific system, its highly non-linear character makes them difficult to solve, even
numerically. Therefore the study of such pattern forming systems is accomplished by
the formulation of phenomenological *generic models* or *model equations*. Such models
must have pattern formation properties as close as possible to those of the original system, but they should be much easier to treat analytically or computationally [Cross &
Hohenberg, 1993; Venkataramani & Ott, 2001]. Examples of model equations are the
complex Ginzburg-Landau equation, the Swift-Hohenberg equation, and the Kuramoto-
Sivashinsky equation. The application of linear stability analysis of surface perturbations

can provide information about the characteristic lengths at early times and small amplitudes, when the non-linear terms influence is not yet important.

Through this work we often refer to the Kuramoto-Sivashinsky (KS) equation [Kuramoto, 1984; Kuramoto & Tsuzuki, 1976; Sivashinsky, 1979], which has been proposed as generic non-linear continuum description for various pattern formation phenomena. The common feature of such a systems is that after an early regime characterized by the emergence of a periodic pattern, the system evolves in a disordered state.

1.4 Universality in pattern formation systems

An important feature of the generic models that describe diverse pattern formation systems is that their properties are thought to be independent of the detailed mechanisms leading to the instabilities. Thus, they are *universal* in the sense that can be applied to diverse physical systems. This work is based on considering that the pattern formation phenomena found in ion beam sputtering, water jet cutting, and laser-induced jet-chemical etching have common elements, and in consequence they can be described by similar phenomenological models.

Ion beam sputtering (IBS) is the removal of material from the surfaces of solids through the impact of energetic ions. The ion beam has a uniform flow density over all the surface and it is directed with a certain fixed incidence angle. It has been found that the evolution of the surface is governed by the competition between the dynamics of erosion and surface diffusion processes. The evolution of the surface is characterized by an early stage where periodic ripples aligned parallel or perpendicular to the bombarding ion beam appear; then for later stages, the surface is roughened (see [Carter, 2001; Valbusa *et al.*, 2002]). The Kuramoto-Sivashinsky(KS) equation and further generalizations have been used to describe such crossover and another features of the experiment (see [Castro *et al.*, 2005; Makeev & Barabási, 2004a,b] and references therein).

Water jet cutting (WJC) consists in to cut a surface with a high velocity jet of water that contains abrasive particles [Momber & Kovacevic, 1998]. With this technique is possible to obtain clean cuts of hard materials. For relatively high feed velocities of the jet, the walls of the cut develop unwanted periodic striation. Although WJC and LJE are structuring techniques based on quite different microscopic mechanisms (mechanical grinding vs. chemical dissolution), we assume that behind the similarity of the patterns there is a universal behavior. For example, based on an experimental observation in WJC that relates the diameter of the water jet with the striation wavelength, a working hypothesis for LJE about the ripple length dependence with the laser power is supported. On the other hand, Friedrich *et al.* have proposed a phenomenological theory to describe the striation phenomena in WJC [Friedrich *et al.*, 2000b; Radons *et al.*, 2004]. In a comoving frame, the dynamics of the cutting front is represented by a generalized Kuramoto-Sivashinsky equation, which takes into account the localized action of the water beam.

1.5 The extended model and the LJE simulation

The formation of surfaces in general is influenced by a large number of processes, and their relevance and mutual interaction is usually unknown. There is experimental evidence that in many cases, the resulting surfaces may develop self-affine properties, which can be described by the concepts of the fractal geometry and theories of scaling and universality [Meakin, 1998]. The kinetic roughening properties of surfaces can be categorized into universality classes by calculating scaling exponents [Barabási & Stanley, 1995; Halping-Healy & Zhang, 1995; Krug, 1997; Ódor, 2004]. Two main approaches have been proposed for the analysis of such problems. The first is the formulation of discrete models based on simplified representations of the physical processes. The second approach is based on stochastic continuum equations, which provide a coarsed-grained description of the interface from which the asymptotic scaling behavior can be characterized [Barabási & Stanley, 1995; Marsili *et al.*, 1996].

For the evolution of the surface in the LJE process, we have opted for an emphasis on the first approach. Discrete models based on simple algorithms have played a central role in surface science from the very beginning of the computer era. These models mimic the essential physics and do not necessarily depend on detailed microscopic knowledge of the processes. In particular, we are interested on *solid-on-solid* (SOS) models where the surface is represented by a lattice and overhangs are no allowed. Processes like deposition or erosion can be represented by the creation or annihilation of cells at random positions on the surface, whereas surface diffusion processes can be represented by hops between neighboring lattice sites. Within the *kinetic Monte Carlo* method such random events are simulated with certain event probabilities [Newman & Barkema, 1999].

For the LJE problem we extended and adapted a 1+1 dimensional SOS model proposed initially for ion-sputtering by Cuerno, Makse, Tommassone, Harrington and Stanley (CMTHS) [Cuerno *et al.*, 1995]. In the CMTHS model ripples appear at early stages of the evolution, and the surface roughens for later stages. The authors demonstrated that the scaling properties of the CMTHS model are close to those of a generic noisy KS equation. Due to the limited feasibility of the original CMTHS model to generate a ripple regime, we propose an *extended model*, which mainly relies on an improved estimation of the local curvature [Mora *et al.*, 2005a,b]. The obtained ripples are more regular and their wavelength is longer, which is a practical requirement for the simulation of the localized etching. We propose the erosion and surface diffusion rules of the extended model to simulate the etching process, in particular its unstable behavior.

In the LJE experiment the processes are mainly determined by the heat resulting from the absorption of the laser beam. Given the thickness of the metallic sample and the relatively large dwell times of the laser beam, the back surface of the sample reaches approximately the same temperature as the front surface on which the radiation is incident. Therefore, using a *thermally thin approximation* a temperature field can be defined. We define the *etching probability distribution* of applying the erosion and diffusion rules to be directly proportional to the estimated temperature field. The laser power and the temperature threshold above which etching can occur define the amplitude and width of the etching probability distribution.

The *LJE simulation* consists in the application of certain etching probability distribution which moves with a *feed velocity* v_f over the lattice that represents the surface. Any portion of the surface is exposed to the extended model rules during a dwell time, which depends on the feed velocity and the diameter of the etching probability distribution. Therefore, the final topography at the bottom of the kerf is certain stage of the temporal evolution of the extended model determined by the dwell time. For a certain parameter set, it is possible to obtain transitions from uniform kerfs to rippled kerfs when the feed velocity or laser power is varied. As in the experiments, above a threshold of feed velocity, the etching front becomes unstable and the kerf features ripples whose wavelength decreases with increasing feed velocity. In the simulations with increasing laser power, the ripple length increases. , this is in qualitative agreement with the experimental observations.

During the analysis of the evolution of the ripple regimes of the CMTHS and extended models we have found a coarsening phenomena. Namely, after the onset of the ripples their wavelength increases in the course of time. This effect plays a main role in the qualitative agreement of the LJE simulations with the experiment. On the other hand, the appearance of coarsening indicates that the behavior of the extended model differs from a pure Kuramoto-Sivashinsky dynamics, where the selected length scale remains constant [Politi & Misbah, 2004].

In order to understand better the analytic properties of the extended model, we have studied its correspondence with a continuum stochastic differential equation. Following a method based on the master equation proposed by Vvedensky and others (see [Chua *et al.*, 2005] and references therein), we show how a modified Kuramoto-Sivashinsky (KS) equation can be associated to the extended model. This equation is a gradient expansion where the low order terms correspond to the KS equation, while higher order terms can be held responsible for the coarsening phenomena. Moreover, we have shown that considering the inhomogeneity caused by the moving etching front, a continuum Kuramoto-Sivashinsky equation in a frame comoving with the laser beam can be associated to the LJE simulation [Mora *et al.*, 2005b].

The main advantages of the discrete *extended model* and the *LJE simulation* are simplicity and versatility. The probabilistic nature of the method allows a minimal but meaningful representation of the essential physical processes and experimental parameters of the LJE technique. Remarkably, the pattern formation features of the model qualitatively resembles the experimental results. This kind of phenomenological models constitute an intermediary between the experiment and the formulation of an analytical theory with predictive power. Based on the simulation results, new experiments can be proposed. Therefore hypotheses about the relationships between the experimental parameters and the properties of the patterns can be evaluated systematically. The ultimate goals are the identification of the mechanism that creates the ripples and the formulation of the corresponding microscopic equations. Eventually, this advanced understanding will provide a means to control the characteristics of the patterns for future technological applications.

1.6 Overview

This work is organized as it follows. We start in Chapter 2 presenting the basics of the experimental setup. We show two type of structures analyzed through this work: uniform and rippled kerfs. Then, we propose a qualitative description of some of the microscopic processes occurring during etching in the LJE technique. We analyze in Chapter

3 the morphology of a series of kerfs structured varying the feed velocity. Using standard Fourier techniques, we are able to distinguish between intrinsic ripple formation and periodic structures resulting from external frequencies. In Chapter 4 we present some advances in the application of a new stochastic method based on a Fokker-Planck equation for the description of the scale-dependent roughness of surfaces submitted to electropolishing. Returning to the LJE experiment, in Chapter 5, we analyze the heat transport problem and justify the use of a the thermally thin approximation in order to estimate of the temperature field.

We review in Chapter 6 some properties of the Kuramoto-Sivashinsky (KS) equation and its application in the description of pattern forming systems. The first application is ion beam sputtering. We present the basic experimental and theoretical facts and how variations of the KS equation have been used to model the temporal evolution of ion sputtered surfaces. As second example, we review a modified KS equation proposed by Friedrich et al. [Friedrich *et al.*, 2000b] to explain striation phenomena in water jet cutting. In order to propose a working hypothesis on the relationship between the ripple length and the average size of the etching front, we analyze there another experimental series where the power is varied.

The novel contributions of the work, namely the *extended model* and *LJE simulation* are presented in Chapters 7 and 8. We start Chapter 7 reviewing the CMTHS model and introducing the *extended model*. Then we perform a comparative analysis of the scaling properties and surface evolution for both models. There we show how the improved estimation of the local curvature of the extended model produces a ripple regime suitable for the simulation. The application of the extended model to simulate the LJE process is presented in Chapter 8. Using the temperature field defined in Chapter 5, we define the etching probability distribution, which is used to simulate the action of the laser beam and etchant jet. The simulated kerfs for varying velocity or varying power show ripple regimes that are in qualitative agreement with the experimental results. The general conclusions are presented in Chapter 9. Preliminary advances in the generalization to 2+1 dimensions of the extended model and the LJE simulation are presented in the appendix.

Chapter 2

The LJE Technique

Since the last decades, the microstructuring of metallic surfaces has been a topic of intensive research due to its current and promising future applications in the precise fabrication of devices for opto-electronics, microsurgery and microfluidics, among many others. In order to avoid unwanted effects produced by high power laser melting and evaporation techniques (i.e. heat degradation, and microstructural changes in the heat affected zone), diverse electrochemical dissolution laser-enhanced techniques have been proposed.

Laser-induced wet chemical etching in semiconductors [Lee *et al.*, 1990; Takai *et al.*, 1988b], ceramics [Lu & Ye, 1996a; Lu *et al.*, 1988; Lu & Ye, 1996b; Takai *et al.*, 1988a], and metals [Datta, 1998a,b; Datta *et al.*, 1987; von Gutfeld & Sheppard, 1998; von Gutfeld *et al.*, 1988] is a well established microstructuring technique. Controlling the quality of final structures is the main concern from an experimental and theoretical point of view. In a variation of the experimental technique developed by Metev, Stephen et. al. (see [Metev *et al.*, 2003; Nowak & Metev, 1996; Stephen *et al.*, 2002, 2004]), and currently implemented by Rabbow et al. [Rabbow *et al.*, 2005], a combination of a focused laser beam and a jet of etchant induces an etching reaction on metallic surfaces producing holes and kerfs (grooves) within a micrometer scale. This technique is called *laser-induced jet-chemical etching* (LJE). In Section 2.1 we present the experimental setup and the main parameters of the experiment. The term "etching front" is defined in Section 2.2. In Section 2.3 we show the features of the diverse obtained structures. A qualitative description of the different physical and chemical processes is presented in Section 2.4.

2.1   The experimental setup

In the experimental setup, foils of stainless steel or titanium are immersed in a solution of etchant based on phosphoric or sulfuric acid. Under normal temperature conditions, the layer in contact with the liquid is *passivated* spontaneously, thus isolating the metallic sample from the etchant action. In the case of stainless steel, the passivation of the surface appears when the environment can provide enough oxygen to form a Cr_2O_3 oxide layer, which prevents further transport of etching species to the underlying metallic surface. In order to structure the surface, a focused laser beam enhanced by a coaxial jet of etchant is directed perpendicularly to the surface. The area below the laser spot is heated, and almost immediately the heat spreads to a wider zone. Above a temperature threshold, the passivation layer is removed and thermally activated chemical etching starts there.

The function of the etchant jet is to increase the etching rates, providing fresh etchant and enhancing the transport of dissolved material and other reaction products. Due to convection, the jet creates a cooling effect, which maintains the temperature field concentrated in a small region. In addition, an external electric potential could be applied to enhance the reaction kinetics and reaction products transport.

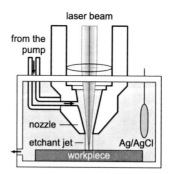

Figure 2.1: (Color online) Schematic diagram of the etching chamber. A focused laser induces an etching reaction enhanced by a coaxially running jet of etchant. An external pump injects etchant into the nozzle, and the resulting jet impinges the workpiece surface. The etchant is ejected out of the chamber through an outlet and returns to the pump to complete the etchant circulation.

2.1.1 The etching chamber

A schematic diagram of the *etching chamber* is shown in Figure 2.1.   Foils of
Fe/Cr18/Ni10 stainless steel or titanium are immersed horizontally in a solution of etchant
based on H_3PO_4 or H_2SO_4. Typical dimensions of the stainless steel samples are 2×2
cm size and a thickness of 200 μm. Inside the chamber, a nozzle ejects the etchant with
velocities up to 2.5 m/s. The nozzle is fed by an external peristaltic pump, which due to
its operation principle, introduces pulsations in the etchant jet in the range of 1-3 Hz. The
influence of these pulsations is analyzed in forthcoming chapters. The chamber also has
an outlet from which the used etchant returns to the pump, which in turn injects it back to
the nozzle, and a constant circulation is established. Due to the permanent dissolution of
metals and other reaction products, the concentration of etchant decrease during the experiment. Usually pure etchant produces high erosion rates and low quality of the structures,
while moderate aged etchant produces more uniform structures [Nowak & Metev, 1996].
This aging effect imposes a limit to the exact reproducibility of the experiment.

The etching reaction in the metallic surface produces an electrochemical potential $E(t)$,
which can be measured with respect to a Ag/AgCl reference electrode, which is immersed
in the etching reservoir (see Figure 2.1). The potential not only indicates the change of
state of the metal surface from passive to active, but also due to its proportionality to the
etching rate, provides important information about the temporal evolution of the process.
It is worthy to note that the potential provides an average value of the etching rate over the
region where etching actually occurs. Therefore, large oscillations of the electrochemical
potential during structuring of kerfs are related with the appearance of ripples inside the
kerf.

2.1.2 The complete experimental setup

Figure 2.2 shows a diagram of the complete experimental setup. The laser source is an
Argon ion laser (Coherent Innova 90), which provides a continuous wave radiation
at 514 nm, with an output power about 1 W (in the TEM_{00} mode). The beam has a
diameter of approximately 1.5 mm. In order to reduce its intrinsic divergence before
the focusing system, the beam is passed through a Kepler telescope, and its diameter is
expanded by a factor of 5. Proceeding in this way, the resulting focused spot size is
minimized [Webb & Jones, 2004]. The focusing system is coupled above the nozzle, and

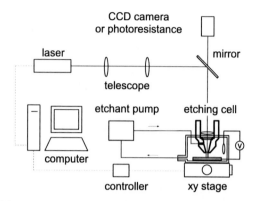

Figure 2.2: (Color online) Schematic diagram of the entire experimental setup. The reaction can be observed with a CCD-camera, and the intensity of the reflected light is measured by photo-resistance. Due to the laser induced etching, the workpiece changes its state locally from passive to active. Changes of the potential are detected with a Ag/AgCl reference electrode.

is isolated from the etchant by a transparent glass. An objective of 50 mm is used as a focusing system, and the obtained spot size is 8 μm. Assuming the gaussian TEM_{00} mode, this means that at 4 μm from the center of the beam, the intensity has fallen to $I_0 e^{-2}$, where I_0 is the maximum value at the center of the beam.

The etching chamber is mounted on a computer-controlled mobile basis (Newport PM-500), which allows the sample to move in the xy plane with respect to the laser beam with *feed velocities* between tenths and hundreds of μm/s. This device provides a precise control of the movements of the sample, but a 100% constant velocity is not achievable. The stage uses a position control technique which compares the output of the system with the desired input and takes corrective action (see a detailed discussion of this issue in [Newport Corporation web page]). In our particular device, this check is performed each 40 μm, and the correcting action modulates the velocity in order to obtain minimal accelerations or decelerations during the control points. As a result, the feed velocity oscillates around a mean value v_m with a frequency $v_m/(40\ \mu\text{m})$. This control mechanism produces periodic structures of 40 μm, which are detectable in most of the analyzed structures. We will discuss the role of this external perturbation on the observed ripple formation in Chapter 6.

The experimental setup is an automatized system which allows for structuring of holes or kerfs by varying the most relevant external parameters: laser power, etchant jet velocity, and feed velocity of the sample. Some details of the process are observed with a CCD camera located above the etching chamber. With a photo-resistance located in the same position, it is possible to measure the intensity of the reflected light by the surface. This measurement only gives an account of reflected light that makes its way through the optical system in the inverse direction of the incident light and reaches the photo-resistance. A lot of light is reflected and scattered in other directions.

The combination of electrochemical potential and reflection measurements can be used for monitoring the etching dynamics. We use this information in Chapter 3 to trace in time the process of formation of a single ripple. However, because these measurements are mean values of processes occurring simultaneously in different points (with different etching rates and temperatures), detailed information about the surface morphology at a microscopic scale can not be inferred from them.

2.2   The etching front

During the structuring of a kerf, various processes occur at different points of the affected surface. Figure 2.3 illustrates a portion of the metallic sample moved with feed velocity v_f relative to the laser beam. The heating process resulting from the absorption of laser radiation spreads over a region much broader than the dimensions of the kerf. Depassivation only occurs above a temperature threshold, which is strongly dependent of the concentration of etchant [Shin & Jeong, 2003]. The *etching front* is defined as the region where the passivation layer has been removed, and the etching reaction occurs. Taking the sample as a reference, the etching front produced by the laser beam drifts through the surface with a speed of magnitude v_f, heating, and subsequently depassivating and etching formerly passivated portions of the surface. Beyond the rearmost part of the advancing etching front, the temperature decreases below the depassivation threshold, and the surface is passivated again. An important physical quantity to be considered is the *dwell time*, which is defined as the time the etching front resides on a determined position of the surface. It can be estimated by

$$t_{\text{dwell}} = \frac{d_{\text{front}}}{v_f},$$ (2.1)

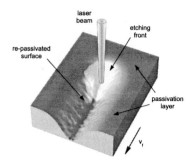

Figure 2.3: Diagram showing the etching front during the structuring of a kerf. The heat produced by the laser absorption depassivates the surroundings of the area illuminated by the laser beam forming the *etching front* where the etching reactions actually occur. The frontier of the etching front is defined by the *depassivation temperature*, above which etching reactions can occur. Regions that are below this temperature are protected by a *passivation layer*. Inside the kerf, beyond the action of the beam, there are zones where the temperature has decreased, and a passivation layer is formed again (*repassivated*).

where d_{front} is the diameter of the etching front.

2.3 Overview of the etched structures

2.3.1 Structuring of holes

For structuring holes on the metallic samples, the control system is programmed to expose the static sample (feed velocity $v_f = 0$) during a short interval of time, depending on the desired depth. Figure 2.4 shows some examples of the obtained holes. In order to analyze the evolution of the structuring of a hole, a series of holes with increasing exposure times has been performed [Stephen, 2004]. Afterwards, a transversal cut has been performed in order to obtain a profile for each hole. Figures 2.4(a)-(c) show optical microscopic images corresponding to the hole profile after 2,6 and 20 seconds. Figure 2.4(d) shows a top view of a typical hole.

Figure 2.4(a) shows how, at early times, the etching process is driven by removal of small pieces of metal. For $t = 6s$, and $t = 20s$ (Figures 2.4(b) and (c)) the profiles show

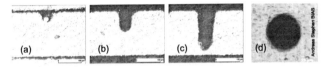

Figure 2.4: Examples of structuring of holes. For laser beam power of 5W: hole profile after (a) 2 seconds, (b) 6 seconds and (c) 20 seconds. (d) Top view image of a typical structured hole. Images cortesy from Andreas Stephen at BIAS-Bremen.

a "U" shaped bottom with almost vertical walls. Due to large incidence angles at the walls, the reflection is strongly favored, and the light can reach the bottom of the hole after multiple reflections [Lee *et al.*, 1990]. If the removal of dissolved material and the supply of fresh etchant are efficient enough, the etching of the bottom can continue at large depths. Eventually, the sample can be drilled through for large exposure times.

2.3.2 Kerfs structured with steady etching front

When the metallic sample is moved with respect to the laser beam axis, diverse kerfs or grooves are obtained. In the case that a steady etching front is formed, the resulting kerf surfaces have small-scale roughness, but on average there are no variations in depth and shape of the walls along the kerf. In what follows, we refer to such structures as *uniform kerfs*. Figures 2.5(a)-(d) illustrate how the transversal profile of the kerf varies with increasing feed velocities v_f. As might be expected, a larger feed velocity mean smaller residence or dwell time, and the kerfs become shallower. The kerf depicted in Figure 2.5(e) has been performed on a titanium sample and illustrates the high aspect ratios that can be obtained with the LJE technique. Figure 2.5(f) shows a top view of two typical uniform kerfs eroded on stainless steel with sulfuric acid.

2.3.3 Kerfs structured with unstable etching front

Under certain set of experimental parameters, the etching front becomes unstable and remarkably periodic structures dominate the topography. Figure 2.6(a) shows a top view of a 5 mm long kerf structured on a stainless steel surface with a scan velocity of 5 μm/s. Although the image is blurry, it is possible to distinguish periodic ripples inside.

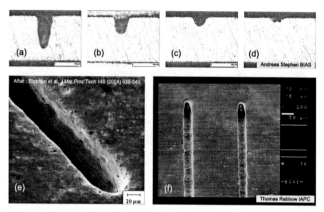

Figure 2.5: Examples of structuring of uniform kerfs. Transversal cut of kerfs for increasing feed velocity: (a) 4μm/s, (b) 10 μm/s, (c) 20 μm/s and (d) 40 μm/s (Experiment and images from Andreas Stephen at BIAS-Bremen). (e) Cut in a 200 μm thick nickel titanium foil (after [Stephen *et al.*, 2004]). (f) Typical uniform kerf structured on stainless steel (Experiment and images from Thomas Rabbow at IAPC-Bremen).

A scanning electronic microscopy (SEM) image shows some details of ripples, which present a "keyhole" shape (Figure 2.6(b)). A single ripple can be described as a rounded cavity, which evolves into a shrunken and shallow trench. At the border of these keyholes, ridges that protrude above the level of the original surface are frequently found. Figure 2.6(c) shows a series of kerfs structured at a constant feed velocity, where ripples with increasing length and width appear for increasing laser power.

2.4 Microscopic description of the LJE processes

We have a limited knowledge of the interaction of diverse processes occurring on a wide range of spatial and temporal scales. The dynamics of laser light absorption, heat, chemical reactions, hydrodynamics and transport phenomena is too complex to be fully modeled analytically or numerically. However, in the following, we propose a simplified scheme to account for the most important processes. Figure 2.7 shows a small portion of the surface that is being etched (small compared to the horizontal dimensions of the effective etching front). Although etching is isotropic, for the sake of simplicity we only represent

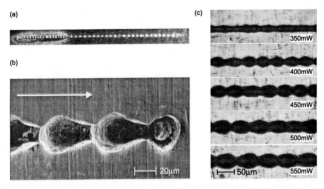

Figure 2.6: Reproduced from Rabbow [Rabbow *et al.*, 2005]: (a) Optical microscope image of a 5 mm long kerf etched on a stainless steel surface with scan velocity 5 μm/s. (b) Scanning electronic microscope (SEM) image of a rippled kerf with the characteristic keyhole shape. The scan direction is indicated by the arrow. (c) Optical microscope images showing the variation of the ripple length for increasing power at constant feed velocity.

the vertical component of the different processes. The diagram depicts the etchant, the bulk of metal and the interface between them where chemical etching occurs.

The etching at a microscopic scale is a complicated process. Once the dissolved material is removed, it leaves behind micro-fractures where fresh etchant comes in, and new dissolution processes occur in a disordered way. Therefore, it can be considered that the etching process occurs within a thin *etching layer*. Within this interface, gradients of concentration of etchant and density of dissolved and non-dissolved metal form a mixture, which is removed progressively by diffusive and convective mechanisms while the etching advances through the material. As a further simplification, let us consider that the removal of dissolved material occurs at the upper part of etching layer, while ionization of the bulk metal occurs in its lower part. In Figure 2.7 we present classification of the processes in three categories: (a) radiation and heat, (b) chemical reactions, and (c) hydrodynamics and transport. They are described as follows:

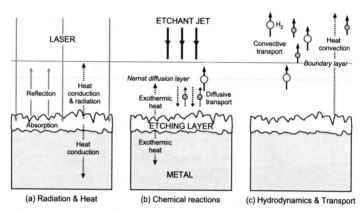

Figure 2.7: Hypothetical representation of the processes occurring during etching, classified according to its nature or origin. (a) Radiation and heat processes resulting from the incoming laser radiation. (b) Chemical etching processes occurring inside the *etching layer*. The fresh etchant and products of reaction are transported within a *Nernst diffusion layer*. Additional heat is produced by the exothermic reaction. (c) Hydrodynamics and transport processes of the etchant jet. The products of the reactions are dissoluted metal (gray circles) and hydrogen molecules (white circles). The surface is also cooled by convective heat transport caused by the circulation of etchant.

2.4.1 Radiation and heat processes

Depending on the polarization, incidence angle and surface roughness, the incoming laser radiation is partially absorbed by the etching layer, as shown in Figure 2.7(a). The reflected light travels back in different directions. The absorbed laser radiation energy is converted into heat enhancing the etching reaction. There is heat conduction from the etching layer to both the etchant liquid and the bulk of the metallic sample. The etching layer also emits heat in the form of electromagnetic radiation. This effect is important when the radiating surface is relatively large.

2.4.1.1 Chemical reactions

In the actual experiment, there are many species involved and various options that can account for the etching reactions. Here we discuss the basic mechanism of etching of stainless steel surfaces by sulfuric acid (the reactions with phosphoric acid are similar). The protons of the sulfuric acid (H_2SO_4) react with the iron, nickel and chromium of the steel, producing hydrogen and dissolution of metal ions. At the atomic level, the simultaneous oxidation-reduction reactions can be described as [Rabbow, 2005]:

- Dissociation of the sulfuric acid:

$$H_2SO_4 \rightarrow 2H^+ + SO_4^{2-} \qquad (2.2)$$

- Oxidation or *anodic* reaction (ionization of the metal):

$$Fe \rightarrow Fe^{2+} + 2e^- \qquad (2.3)$$

- Reduction or *cathodic* reaction (formation of hydrogen): The protons from the acid combine with the electrons released by the metal to form hydrogen radicals, and later, hydrogen molecules.

$$2H^+ + 2e^- \rightarrow H_2 \uparrow \qquad (2.4)$$

These molecules aggregate themselves to form gas bubbles, which drift upwards due to buoyancy, stirring and convection.

- The desorpted metallic ions can combine with the negative ions of the acid to form salts, in this case Ferrous sulfate:

$$Fe^{2+} + SO_4^{-2} \rightarrow FeSO_4 \tag{2.5}$$

- The sum reaction:

$$Fe + H_2SO_4 \rightarrow FeSO_4 + H_2 \uparrow \tag{2.6}$$

with similar reactions for the ionization of the nickel and chromium. In this version of the experiment, there is no applied external potential, but from the electrochemical point of view, the system consists of simultaneous anodic and cathodic reactions taking place all over the etching front [Datta *et al.*, 1987].

The layer of solution in contact with the upper surface of the etching layer develops a concentration gradient, which ranges from zero on the interface to the value of bulk volume of the incoming etchant. This is called the *Nernst diffusion layer* (NDL), and within its thickness δ, the transport of ions of etchant and products of the reaction occurs exclusively by diffusion, limiting the etching reaction [Bard & Faulkner, 1980]. The thickness of this layer depends, among others, on the concentration of etchant and hydrodynamic conditions above it, and its superior limit is illustrated in Figure 2.4 with a horizontal slashed line. Outside this layer, convective transport maintains the concentration uniform at the bulk concentration.

The etching reaction is exothermic and additional heat conduction occurs from the etching layer to both the liquid etchant and the bulk material (see Figure 2.7(b)). The rate of thermally activated reactions k is exponentially dependent on the temperature T according to the Arrhenius law $k = Ae^{\frac{-E_a}{RT}}$, where E_a is the activation energy and A is a reaction specific factor. Due to this dependency, conditions for a thermal runaway can eventually occur, resulting in a strong increase of the etching and dissolution rates [Anisimov & Khokhlov, 1995; Frank-Kamenetskii, 1969]. This thermal runaway can be proposed as the microscopical mechanism that produces rippled kerfs.

2.4.1.2   Hydrodynamics and transport processes

The microstructuring depends strongly on how fast fresh reactants reach the etching front and how efficiently dissolved material is driven out of the kerf. In Figures 2.7(b) and (c), how the etchant jet perform these transport processes is depicted. In our simplified image, the products of reaction are dissoluted metal and hydrogen gas bubbles (symbolized in the figures by gray and white circles respectively). It is worth noting that, as just another flow running over a surface, a *boundary layer* is created in which the velocity changes from zero at the surface to the free stream value away from the surface [Carslaw & Jaeger, 1959] (the limit is indicated in Figure 2.7 with a horizontal continuous line).

The transport of dissolved metal occurs in two steps: diffusive transport within the Nernst diffusion layer and convective transport above the boundary layer. The relative positions and sizes of the Nernst diffusion layer and the hydrodynamic boundary layer are not known for this experiment, and probably diffusive and convective transport are mixed in some regions. The etchant jet reduces the thickness of the Nernst diffusion layer and in consequence, decreases the time that molecules of fresh etchant and dissoluted metal travel through it, enhancing the etching rate. For the LJE technique, it has been found that for low powers the depth of the holes reach a saturation value, which can be attributed to inefficient transport of reactant and products. Therefore the etching reaction can be considered limited by the mass transport.

Apart from its role in the transport of chemical species, the etchant jet also has the function of cooling the surface by taking away heat due to convection. This is the key feature for the high resolution of the technique: the laser beam provides high temperatures and the etchant jet feeds the reaction and minimizes the etching front size by means of the cooling effect.

2.5   Discussion

We have presented the basics of the LJE technique and an overview of the obtained structures. The heat originated by the laser absorption spreads over a region larger than the laser beam or the kerf dimensions. This spreading is controlled by the cooling effect of the etchant jet. Above a temperature threshold, the passivation layer is removed, and thermally activated chemical etching starts there, forming the *etching front*, which structures the kerf on the surface at velocity v_f.

In most cases, the etching front is steady and the obtained kerf has, apart from small fluctuations, a defined shape with almost constant width and depth. These types of structures are called uniform kerfs. The bottom and walls of the kerfs are always rough at the micrometer scale due to the stochastic character of the etching reactions. On the other hand, for certain parameters ranges of feed velocity, laser beam power or etchant jet velocity, oscillations of the etching front produce kerfs with periodic ripples.

We separately described the different radiation, heat, chemical, hydrodynamic and transport processes. However, a faithful microscopical description is not possible due to the limited information that can be obtained from the experiment. The rates of the etching reactions are not determined only by its own kinetics but also by the diffusive and convective transport of fresh etchant and reaction products. In addition, for a full analytical or numerical analysis, one should consider the heat transport problem in its three forms: convection, conduction and radiation. Certainly the equations and boundary conditions describing all these highly interrelated processes are not known.

Chapter 3

Morphology of the kerfs

Electron and optic microscope images show rich details of the surface of structured kerfs. Although they give insights about the processes involved they do not provide quantitative information about the topography. For such quantitative characterization, we use profilometer measurements which can be correlated with electrochemical potential and reflection measurements.

In this chapter we analyze the topography of a series of kerfs structured with varying feed velocity. From profilometer raw data unwanted trends and misalignments have to be removed. This is necessary because some analysis techniques require profiles with certain degree of global stationarity. From standard Fourier transform analysis, we define the power spectral density (PSD) which is used frequently in the forthcoming chapters.

The kerfs present a range of feed velocities where ripples appear. The ripple length decreases with increasing feed velocity. We investigate if these ripples lengths can be associated with external influences like oscillations of the xy-stage or pulsations of the etchant jet. Finally, the process of formation of a single ripple is investigated comparing the profilometer data with the corresponding electrochemical potential and reflection time series.

3.1 Profilometer measurements

The profilometer performs an adaptive scanning of the structured surface with a low power laser beam. By means of permanent adjustment of the beam focus, the surface heights are

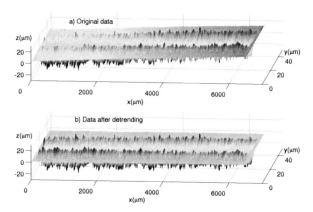

Figure 3.1: Detrending process of typical raw data obtained by the profilometer. (a) The profilometer measurement shows a surface with a trend. (b) The
same surface after a detrending process based on a polynomial regression of
each profile parallel to the xz-plane. All the xz-profiles were detrended using
non-etched xz-profiles as reference.

measured with a resolution of 0.1 μm and the values are stored by the computerized system in a matrix. The lateral resolution in the xy-plane is 1 μm. The raw data usually
includes trends originated from an inherent curvature of the samples. The samples are
curved due to rolling during the fabrication process. The mechanical cutting of the samples can introduce an additional deviation from flatness. A typical measurement surface
data set with such trend is presented in Figure 3.1(a). Note that the surface profiles parallel to the xz-plane are a slightly curved and the height values increase with x. This
particular kerf was structured with scan velocity $v_f = 7$ μm/s and belongs to the series of
kerfs analyzed in Section 3.2. The surface presents ripples with characteristic length of
~ 44 μm but they can not be identified in the figure because the chosen aspect radio.

A detrending procedure is necessary to analyze the topography at small scale by means
of Fourier analysis. Otherwise, long wavelengths will be dominant, concealing the short
wavelength features of the surface. In order to extract the trend, a polynomial least squares
regression of second order is applied to each profile parallel to the xz-plane. Profiles that
lie inside the kerf are detrended interpolating the regression parameters used for profiles
of non etched regions. The resulting surface is shown in Figure 3.1(b). Note that the
plane defined by portions of the surface that have not been etched is now parallel to the

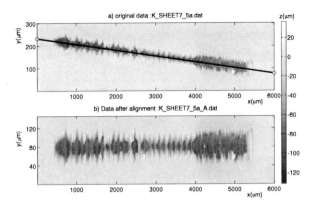

Figure 3.2: Alignment of surface data (the surface is observed from above and the scale of gray levels at the right represents the height z values). (a) The original profilometer surface data (301×6001 points separated 1 μm in each direction) shows a kerf with its axis (black line with circles in both ends) tilted with respect to the x-axis of the data matrix. (b) The surface data after alignment (146×6001 points).

xy-plane.

After detrending, the data frequently presents misalignment, which stems from the fact that the kerf axis does not coincide exactly with the direction used in the profilometer scan. Aligned profiles allow to analyze separately distinct features in the bottoms or walls. In order to illustrate the method, Figure 3.2(a) shows a typical set of surface data with the kerf axis tilted with respect to the x-axis (considering the real aspect ratio, the tilt angle is very small). This particular kerf was structured on a stainless steel sample with 1.9M H_2SO_4 etchant, laser power $4W$ and feed velocity of 6 μm/s. The black line is an estimation of the direction of the main axis of the kerf. The alignment process consists in creating a new matrix with profiles parallel to the y-axis which has been trimmed in both ends and then shifted in a progressive way. Figure 3.2(a) shows the aligned kerf parallel to the x-axis.

The detrending and alignment procedures introduce inevitably spurious information. Regarding the 1 μm lateral resolution, this introduces additional uncertainty to the conclusions that can be drawn about structures of the order of few micrometers. However, because the kerfs analyzed later in Section 3.2 present periodic structures of the order of 40

μm and more, the profilometer measurements are adequate to analyze them. To explore scales below 10 μm, electrochemical potential $E(t)$ and reflection $\phi(t)$ series can been used because they provide better resolution and longer data sets.

3.1.1 Power spectral density

Fourier transform analysis is the natural choice for obtaining information about periodic oscillations in the data. The *power spectral density* (PSD) estimated by taking the modulus-squared of the discrete Fourier transform is used to identify the frequency and wavenumber components. Following the notation used in [Press *et al.*, 1992], if any of the three kinds of data series (profilometer, electrochemical potential or reflection measurements) is represented by the variable h_j where $j = 0, \ldots N - 1$, being N the number of points, its corresponding discrete Fourier transform components are

$$H_n = \sum_{j=0}^{N-1} h_j e^{2\pi i j n/N} \tag{3.1}$$

where $n = -N/2, \ldots, N/2$. In the case of temporal series the related frequencies are :

$$f_n = \frac{n}{N\Delta} \tag{3.2}$$

Where Δ is the sampling temporal interval. For our purposes, only positive frequencies are considered.

The power spectral density (PSD) is defined as

$$\text{PSD}(f) = \sum_{n=0}^{N/2} |H_n|^2 \tag{3.3}$$

If the series is a spatial one, like in the profilometer data, the frequencies f_n are replaced by the wavenumbers k_n, then Δ is the spatial sampling interval and the power spectral density is a function of the wavenumber: $\text{PSD}(k)$. Inverting the values of the wavenumber, one obtains a length axis and dominant peaks can be related directly with characteristic lengths.

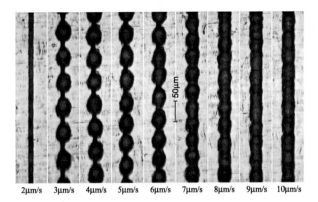

2μm/s   3μm/s   4μm/s   5μm/s   6μm/s   7μm/s   8μm/s   9μm/s   10μm/s

Figure 3.3: Optical microscope images showing a series of kerfs structured with increasing feed velocities v_f from 2 μm/s to 10 μm/s (etchant 5M H_3PO_4, laser power 450mW, etchant jet velocity $v_j = 1.90$ m/s (After [Rabbow *et al.*, 2005]).

3.2   Analysis of a ripple regime

Here we analyze a series of kerfs performed varying the feed velocity v_f. Stainless steel samples were immersed in a solution of 5M H_3PO_4 and exposed to laser power of 450mW. The etchant jet velocity was 1.59 m/s and the feed velocity ranged from 2 μm/s to 10 μm/s, see Figure 3.3. The main feature is the appearance of ripples in the interval of velocities from 3 μm/s to 8 μm/s. The ripples show a complex topography and their characteristic length decrease with the feed velocity. For each kerf, the lengths of single ripples fluctuate perceptibly around the characteristic value.

For this series of experiments, the electrochemical potential $E(t)$ and the intensity of reflected light $\phi(t)$ measurements are more convenient for identifying dominant frequencies. This is due to the fact that $E(t)$ and $\phi(t)$ are measured each 0.01s and the total number of data points is much larger than for the profilometer case. Each kerf is structured during approximately 15 minutes, then the total number of points is approximately 9×10^4. This is almost one order of magnitude larger than the largest profilometer data series (for 10 μm/s approximately 10^4 points see Figure 3.9(a)).

Figure 3.4 shows the value of the ripple lengths (indicated with ○ symbols) as a function of v_f computed identifying dominant frequencies in the PSD on both $E(t)$ and $\phi(t)$ data

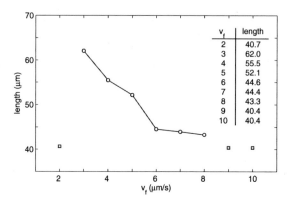

Figure 3.4: Dependence of the ripple length (○ symbols) on feed velocity v_f of the kerfs shown in figure 3.3. The ripple length is estimated from the power spectral density of the electrochemical potential and light reflection temporal series. For feed velocities $v_f = 2$, 9 and 10 μm/s the ripple length (□ symbols) is considered to be related to the undulation of 40 μm produced by the xy-stage.

series (not shown). The fact that the ripple length decreases with the increasing feed velocity indicates that the ripples are of intrinsic nature. If the ripples were triggered by external oscillations like variations on the laser beam intensity or etchant jet velocity, the ripple length would increase proportionally to the feed velocity v_f. We discuss this issue more detailed in Chapter 6.

Kerfs corresponding to $v_f = 2$, 9 and 10 μm/s are product of approximately uniform etching fronts. In the case of $v_f = 9$ and 10 μm/s, the topographies of the kerfs are very rough, see for example Figure 3.9(b). In spite of the roughness of these three kerfs, it is possible to identify a characteristic length of 40 μm (indicated in Figure 3.4 with □ symbols). This length is considered to be related to the oscillation of the xy-stage discussed in Section 2.1.2. As we will see later, this 40 μm length can be observed obiquitously in the data.

Although the 1 μm lateral resolution of the profilometer measurements can not provide detailed information about the internal structure of a ripple, in what follows we analyze the topographies and frequency components of four of the kerfs of Figure 3.3. This topography analysis is necessary, because in principle electropotential or reflection data

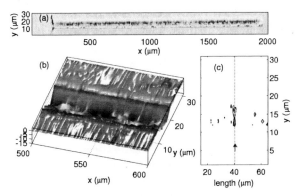

Figure 3.5: Fourier analysis of the bottom of a kerf structured with feed velocity 2 μm/s. Laser power: 450mW, jet velocity 1.59m/s, etchant solution : 5M H_3PO_4. (a) Top view of the kerf surface. (b) Detail of the 3D surface topography. (c) Contour lines of the power spectral density as a function of the length l and the coordinate y: $PSD(l, y)$. The small arrow is located at the corresponding length 40.7 μm estimated in Figure 3.4. The dotted vertical line indicates the 40.3 μm length associated with an oscillation of the xy-stage.

measure mean values proportional to etching rates and do not necesarilly correspond to real surface features at small scales.

In order to study periodic structures along the longitudinal direction of the kerf, we calculated the power spectral density $PSD(k)$ for each profile parallel to the x-axis. Inverting the wavenumbers k, the corresponding lengths l are obtained, leading to the power spectral density as a function of the length $PSD(l)$. Scanning along the coordinate y, and collecting all the corresponding $PSD(l)$, we define a bidimensional power spectral density as a function of the length l and coordinate y: $PSD(l, y)$. This matrix provides information about the periodic features on different parts of the kerf like the walls or bottoms. The $PSD(l, y)$ matrix can be represented as a surface in the three dimensional space. Although this surface is highly irregular, the computation of its contour curves for the largest values of the $PSD(l, y)$ permits to identify dominant lengths as a function of the coordinate y.

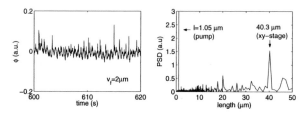

Figure 3.6: (a) Temporal evolution of reflected light $\phi(t)$ measured during the structuring of the kerf of Figure 3.5. (b) Corresponding power spectral density as a function of the length. The peak at 1.05 μm is related to the 2 Hz oscillation introduced by the peristaltic pump. The peak at 40.3 μm is caused by the xy-stage.

3.2.1   Feed velocity $v_f = 2\,\mu\mathrm{m/s}$

The procedure described above is illustrated in Figure 3.5 for a kerf structured with feed velocity $v_f = 2\ \mu$m/s. Details of a portion of the kerf appear in Figure 3.5(b), where an additional rendering and illuminating procedure has been applied in order to emphasize some topographic features. The kerf surface is irregular but can be considered as resulting from an etching front approximately uniform. In Figure 3.5(c) contour curves of largest values of the $\mathrm{PSD}(l,y)$ are centered around 40 μm. The arrow is located at the length estimated in Figure 3.4. The dotted vertical line shows the 40.3μm length associated with the external oscillation caused by the xy-stage.

For these experiments, the peristaltic pump which ejects the etchant jet was operating at a frequency of $\nu = 2$ Hz. If a periodic structure appears in the kerf corresponding to this frequency, its wavelength should be $l' = v_f/2$ Hz. Figure 3.6(a) shows an interval of 20 seconds of the temporal evolution of reflected light $\phi(t)$. Considering the value of the feed velocity, it is possible to obtain the power spectral density as a function of the corresponding length. Proceeding in this way, Figure 3.6(b) shows two peaks corresponding to 1.05 μm and 40.3 μm. The 1.05 μm peak is related with the frequency of the pump: $l' = v_f/\nu = (2\ \mu\mathrm{m/s})/(2\ \mathrm{Hz}) = 1\ \mu$m. The 40.3 μm peak corresponds to the xy-stage oscillation.

The fact that oscillations related with the pump can be identified in the reflection measurements does not necessarily mean that corresponding periodic structures are present in

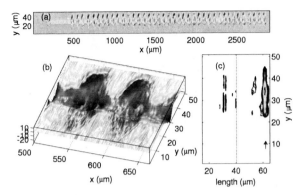

Figure 3.7: Fourier analysis of the bottom of a kerf structured with feed velocity 3μm/s. (a) Top view of the kerf surface. (b) Detail of the 3D surface topography. (c) Contour lines of the power spectral density as a function of the length l and the coordinate y: $\text{PSD}(l, y)$. The small arrow is located at the ripple length of 62 μm estimated in Figure 3.4. The dotted vertical line indicates the 40.3 μm length associated with an oscillation of the xy-stage.

the surface topography. The lateral resolution of the profilometer is too coarse compared with those eventual structures and therefore it is not possible to detect them.

3.2.2 Feed velocity $v_f = 3\,\mu\text{m/s}$

For $v_f = 3\ \mu$m/s, ripples with a complex topography appear (see figure 3.7). The ripples resemble processes of thermal runaways which reach broader areas than in the stationary etching front case, see for example [Anisimov & Khokhlov, 1995; Frank-Kamenetskii, 1969]. An interesting feature is the existence of peaks that protrude above the initial surface level. One of these peaks is located approximately at coordinates $(x = 600\ \mu\text{m}, y = 40\ \mu\text{m})$ in Figure 3.7(b). This implies redeposition or resolidification processes of material on the kerf surface. The contour curves of $\text{PSD}(l, y)$ around 62 μm shown in 3.7(a) coincide with the value computed in Figure 3.4. The contour curves around 40.3 μm show that the perturbation of the xy-mechanism is present in spite of the relatively large contribution of the Fourier components of the ripples. There are also contour curves around 31 μm which is the half of the ripple length 62 μm. This first harmonic is related to the fact that some ripples show two hollow within one period.

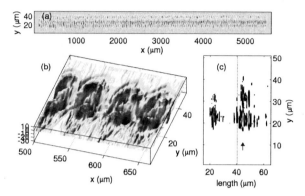

Figure 3.8: Fourier analysis of the bottom of a kerf structured with feed velocity 6 μm/s. (a) Top view of the kerf surface. (b) Detail of the 3D surface topography. (c) Contour lines of the power spectral density as a function of the length l and the coordinate y: $\mathrm{PSD}(l, y)$. The small arrow is located at the ripple length of 44.6 μm estimated in Figure 3.4. The dotted vertical line indicates the 40.3 μm length associated with an oscillation of the xy-stage.

This particular kerf shows very clearly something that can be also noted in other kerfs with ripples: the protruding peaks are always bigger on one side of the kerf. According to Lu and collaborators [Lu *et al.*, 1994], this asymmetry in the topography is expected in all absorbing materials and it is an effect of the polatization of the laser beam. When the polarization plane of the incident beam makes an angle with the scanning direction, a portion of the beam on one side from the center is dominantly p-polarized while the other is dominantly s-polarized. Because the amount of absorbed light depends of the state of polarization, the absorption of light within the spot and the resulting heat transport are not homogeneous, creating asymmetric kerfs.

3.2.3   Feed velocity $v_f = 6\,\mu\mathrm{m/s}$

Another kerf with ripples is analyzed in Figure 3.8 corresponding to feed velocity $v_f = 6\,\mu$m/s. The ripple length is 44.6 μm and a first harmonic around 22 μm appear like in the previous case. Note that again the internal structure of the ripple is completely irregular and there are peaks that protrude above the initial level of the surface at $y \sim 20\,\mu$m.

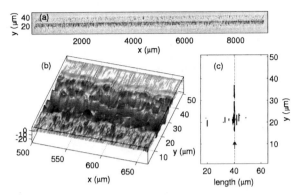

Figure 3.9: Fourier analysis of the bottom of a kerf structured with feed velocity 10 μm/s. (a) Top view of the kerf surface. (b) Detail of the 3D surface topography. (c) Contour lines of the power spectral density as a function of the length l and the coordinate y: $\mathrm{PSD}(l, y)$. The small arrow is located at the ripple length of 40.4 μm estimated in Figure 3.4. The dotted vertical line indicates the 40.3 μm length associated with an oscillation of the xy-stage.

3.2.4 Feed velocity $v_f = 10\ \mu\mathrm{m/s}$

When the feed velocity increases, the kerfs become shallower and periodic ripples disappear. This is illustrated in Figure 3.9 for $v_f = 10$ μm/s where a highly rough topography is obtained. Nevertheless, the 40 μm length related with the xy-stage can be identified in the Fourier analysis, see Figure 3.9(c). For the purpose of detecting the pump oscillation in the temporal data, Figure 3.10(a) shows a portion of reflection data for feed velocity $v_f = 10$ μm/s. In the $\mathrm{PSD}(l)$ of the Figure 3.10(b), two peaks corresponding to 5.1 μm and 40.3 μm appear. The 5.1 μm peak is related with the frequency of the pump: $v_f/\nu = (10\ \mu\mathrm{m/s})/(2\ Hz) = 5\ \mu$m. Like in the case discussed in the figure 3.6, it is not possible to confirm unambiguously the existence of such periodic structures in the profilometer data.

3.3 Tracing ripple formation

Available experimental measurements can provide some insight into the understanding of the process of creation of a single ripple. For this purpose, we correlated the profilometer

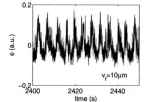 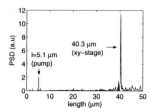

Figure 3.10: (a) Temporal evolution of reflection of light measured during the structuring of the kerf of Figure 3.9 with $v_f = 10\,\mu$m . (b) Corresponding power spectral density as a function of the length. The peak at 5.1 μ results from the 2 Hz oscillation introduced by the peristaltic pump. The peak at 40.3 μm is introduced by the xy-stage.

measurements with the electrochemical potential $E(t)$ and the intensity of reflected light time series $\phi(t)$, see figure 3.11. The top view of a portion of the kerf surface is represented in Figure 3.11 showing after $x \sim 720\,\mu$m, two consecutive ripples. The height of the surface is represented in gray scale, where white regions correspond to highest values and black regions represent lowest values. The intermediate gray represents regions which are close to the original level of the surface. The depicted contour lines are a guide to the eye and emphasize the topographic features. Note that the white regions located approximately at $x \sim 745\,\mu$m and $x \sim 795\,\mu$m correspond to peaks that protrude above the initial surface level.

The electrochemical potential $E(t)$ shown in figure 3.11(b) oscillates between two basic stages (note that the potential has negative values): i) large negative values (between (-0.15,-0.14)) which look roughly like plateaus and correspond to low etching rates; and ii) small negative values (below -0.16) which look like hollows and correspond to higher etching rates. By comparing the figures (a) and (b) it is evident that the cavities inside the ripples correspond to the large etching rates and the "bridge" that connects of one ripple with the next one corresponds to small etching rates. Note that there is a height maximum at position $x \sim 780\,\mu$m, which is close to the level of the original surface and can be related to a maximum at $t \sim 1195$s in the middle of the plateaus of the electrochemical potential series.

In Figure 3.11(c), the intensity of reflected light $\phi(t)$ can be also correlated with the surface and potential measurements. High etching rates correspond to high absorption of light and thus low intensity of reflected light values. On the other hand, just after the

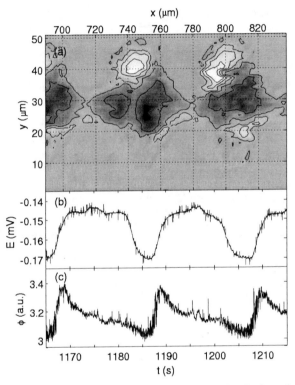

Figure 3.11: Comparison of different measurements during the formation of ripples for a kerf structured with feed velocity 3 μm/s. (a) Top view of a portion of the kerf surface measured with the profilometer. (b) Corresponding electrochemical potential. (c) Corresponding reflected laser light.

cavity of the ripple is formed, when the "bridge" between the ripples starts to form, then absorption of light decreases. This explains the strong increase of the reflection occurring between $t = 1185$s and $t = 1190$s.

3.4 Discussion

Profilometer measurements together with electropotential and reflection data provides valuable information about the kerfs' topography. However, low resolution, mandatory pre-processing techniques and the intrinsic stochasticity of the processes impose limits to the quantitative analysis. We characterized a series of kerfs performed varying the feed velocity. For a range of feed velocities periodic ripples appear. The main feature is the decrease of the ripple length with the increase of the feed velocity. Outside this ripple regime, kerfs resulting from a approximately uniform etching front present rough surfaces.

An undulation of 40.3 μm related with the controlling mechanism of the xy-stage is clearly noticeable in kerfs with or without ripples. The 2 Hz oscillations of the etchant flow generated by the peristaltic pump can be only distinguished in the electrochemical potential and the reflection data due to the limited resolution of the profilometer measurements. Eventually, corresponding periodic structures ranging from 1 μm to 5 μm could be found in the surface. However, due to the limits of resolution, these lengths can not been found in profilometer data. The differentiation between ripples originated from intrinsic instabilities and those triggered by external frequencies is important for the understanding and modelling presented in the next chapters.

Chapter 4

Roughness analysis of electropolished surfaces

Attempting to model the processes resulting in small-scale roughness of surfaces and to compare with experimental measurements calls for numerical methods that allow a quantitative characterization being as complete as possible. In this chapter we summarize our previous work on the characterization of the roughness of surfaces produced by electropolishing of metallic samples [Haase *et al.*, 2004; Mora & Haase, 2004a,b; Mora *et al.*, 2004]. The electropolishing technique and the LJE experiment have some common elements at the microscopic level, particularly in the dynamics of the electrochemical reactions. However, due to fundamental differences between the experimental setups, the physical processes involved in both systems are barely related and no comparisons can be drawn. The purpose of this chapter is illustrate the methods of roughness analysis taking the profilometer data of the electropolished surfaces as example. Eventually the same analysis could have been performed on the rough surfaces of the bottom of the uniform kerfs structured in the LJE experiment, but it was not possible due to the relative scarcity of data and the existence of a periodicity 40.3μm produced by the controlling mechanism of the xy-stage.

(a)   (b)

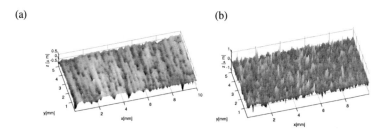

Figure 4.1: Surface topography of brass surfaces after electropolishing: (a) using methanol-electrolyte. (b) Using glycerine-electrolyte.

4.1   Electropolishing of metallic surfaces

Electropolishing is a technical process to obtain smooth and shiny surfaces by means of electrochemical removal of metal. Two samples of brass are positioned face to face and vertically into the electrolyte. The samples are connected to a DC power supply working as the anode and cathode pair of the electrochemical system. The complex surface structures are formed through the interplay of two phenomena: (i) isotropic roughening due dissolution of metal and formation of oxygen bubbles, and (ii) an anisotropic structuring that occurs when a falling film of expended electrolyte containing dissoluted metal interacts with the rising gas bubbles released from the metal surface by buoyancy. As result, the surface is smoothened, but depending on the type of electrolyte, a characteristic pattern of vertical channels or lines at the micrometer scale can be formed.

Gerlach [Gerlach, 2002; Gerlach *et al.*, 2004] obtained electropolished brass samples with two different electrolyte solutions (among others) containing phosphoric acid and and amount of alcohol, denoted as *methanol-electrolyte* and *glycerine-electrolyte*. The surface topography is measured with a laser-assisted profilometer (See Figure 4.1). The surface electropolished with the methanol-electrolyte is very smooth and shiny with tiny ripples in the vertical direction. On the other hand, the surface treated with glycerine-electrolyte has high peaks, it is rougher and dull and suggests self-affinity. Figures 4.2(a,c) display the characteristic height profiles $z(x)$ for both electrolytes. In contrast with the large scales, a zoom into the methanol-electrolyte characteristic profile shows that at small scales the surface is rougher than in the glycerine-electrolyte case [see Figures 4.2(b,d)].

By means of the power spectral density (PSD) (see Chapter 3), two scaling regions for each surface can be identified (see Figure 4.3). Our interpretation is that the large scales

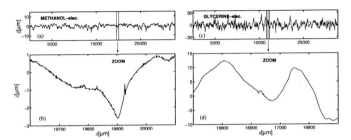

Figure 4.2: For methanol-electrolyte: (a) Characteristic profile and (b) zoomed region. For glycerine-electrolyte : (c) Characteristic profile and (d) zoomed region.

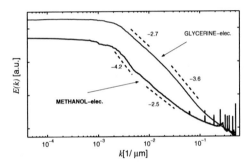

Figure 4.3: Power spectral densities $E(k)$ for methanol-electrolyte and glycerine-electrolyte

(low wave numbers) provide information about the process of formation of gas lines. The scaling region at small scales (high wave numbers) should correspond to the elementary electrochemical processes that occurs independently of the combined action of the falling fill and rising gas bubbles.

4.2 Wavelet analysis

The power spectral density can only provide estimates of the global roughness and gives limited information about the monofractal properties of the profiles. In order to explore the scaling properties of the surface roughness we have applied recently developed

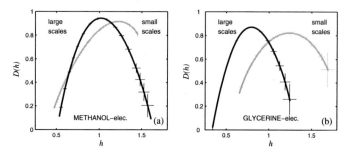

Figure 4.4: Spectra of Hölder exponents $D(h)$ for (a) methanol-electrolyte, and (b) glycerine-electrolyte for two different scaling regions.

wavelet techniques [Mallat, 2001]. Wavelets are defined as two parameter families of functions that are localized in space and scale and can be used to unfold the hierarchical structure of the analyzed signal. The continuous wavelet transform (CWT):

$$W_\psi f(a,b) = \frac{1}{a} \int\limits_{-\infty}^{+\infty} f(x) \, \overline{\psi\left(\frac{x-b}{a}\right)} \mathrm{d}x \qquad (a,b \in R, \, a > 0) \tag{4.1}$$

decomposes the function $f(x)$ hierarchically in terms of elementary components $\psi[(x-b)/a]$, which are obtained from a single *mother wavelet* $\psi(x)$ by dilations and translations. Here, $\overline{\psi}(x)$ denotes the complex conjugate of $\psi(x)$, a the scale and b the shift (space) parameter.

While the Continuous Wavelet Transform (CWT) provides highly redundant information, the wavelet transform maxima lines contain the essential information about the evolution of scaling properties of irregularities across scales. Thus, they can be considered as a fingerprint of the signal [Haase & Widjajakusuma, 2003; Haase *et al.*, 2002]. The Wavelet Transform Modulus Maxima (WTMM) method allows to extract the scaling characteristics of irregular functions, which may even follow different power laws in different regions [Muzy *et al.*, 1994]. The method is a generalization of the classical multifractal formalism. Instead of relying on box-counting techniques, wavelets are introduced as oscillating variants of box functions. The singularity of a function at a point x_0 can be characterized by the Hölder exponent h, which can be considered as extension of the "order of differentiability" to non-integers. The higher the strength of the singularity the lower Hölder exponent. For multifractal functions, the singularity spectra $D(h)$ measures the global

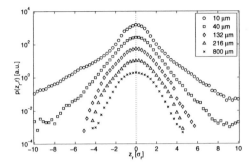

Figure 4.5: Probability density functions of the height increments z_r at different scales (symbols) for the surface electropolished with methanol-electrolyte.

distribution of singularities having different Hölder exponents and can be calculated by means of the WTMM method [Mallat, 2001; Muzy *et al.*, 1994].

To characterize the multifractal behavior of the two electropolished surfaces, we have calculated for the two identified scaling regions the corresponding distribution of Hölder exponents. The spectra of the Hölder exponents displayed in Figure 4.4 show similarities at the small scales for the two electrolytes. The most frequently found singularity corresponds to the Hölder exponent h_0 with the highest value of $D(h_0)$. In both cases this value is around 1.25. $D(h_0)$ is smaller for glycerine-electrolyte indicating that the corresponding singularities are less frequent. For large scales, however, the glycerine-electrolyte spectrum is shifted to smaller h-values which corroborates the observation, that the glycerine-electrolyte produces sharper high peaks on large scales than methanol-electrolyte The singularity spectra $D(h)$ can therefore be considered as a quantification of the qualitative conclusions about roughness at different scales drawn from inspecting the profiles.

4.3 Stochastic analysis

The multifractal characterization is still incomplete, since only 2-point correlations are involved in this formulation. In order to get insight of the possible correlations of the roughness measures on different scales, it is possible to consider the scale dependence of surface roughness as a stochastic process. It is usual to characterize the complexity of a rough surface by the statistics of the height increment:

$$z_r(x) = z(x + r/2) - z(x - r/2) \tag{4.2}$$

depending on the scale r. The probability density function (PDF) of the height increments z_r for various scales r are plotted in Figure 4.5. In order to simplify the notation of the height increments we define $z_1 \equiv z_{r_1}, z_2 \equiv z_{r_2}, \ldots$ an so on. The PDFs are normalized to their respective standard deviations σ_r, and shifted in vertical direction for clarity. For small scales, the shapes of the curves deviate strongly from Gaussian distributions indicating pronounced intermittency effects. For large scales the shape of the PDFs is approximately Gaussian.

4.4   The Markovian description of the roughness

In a series of papers (see [Friedrich *et al.*, 2000b; Renner *et al.*, 2001] and references therein), a new approach for the stochastic analysis has been proposed to reconstruct equations of the underlying stochastic process from experimental data, provided the process is Markovian. This stochastic approach turns out to be a promising tool also for the other systems with scale dependent complexity as turbulence [Renner *et al.*, 2001; Tutkun & Mydlarski, 2004], financial data [Friedrich *et al.*, 2000a], analysis of surface roughness [Wächter *et al.*, 2004; Waechter *et al.*, 2003], and reconstruction of surfaces [Jafari *et al.*, 2003]. The method has been also applied recently to the analysis of the stochasticity of the North Atlantic Oscillation(NAO) index [Collette & Ausloos, 2004; Lind *et al.*, 2005].

Complete information about the stochastic process would be available from the knowledge of all possible n−scale joint probability density functions (PDF) $p(z_1, r_1; z_2, r_2; \ldots; z_n, r_n)$, which describes the probability of finding simultaneously the increments z_1 on the scale r_1, z_2 on the scale r_2, and so forth up to z_n on the scale r_n. The n−scale joint PDF can be expressed by multiconditional PDFs:

$$\begin{aligned}
p(z_1, r_1; \ldots; z_n, r_n) &= p(z_1, r_1 | z_2, r_2; \ldots; z_n, r_n) \cdot p(z_2, r_2 | z_3, r_3; \ldots; z_n, r_n) \\
&\quad \cdot \ldots \cdot p(z_{n-1}, r_{n-1} | z_n, r_n) \cdot p(z_n, r_n)
\end{aligned} \tag{4.3}$$

where the conditional PDF $p(z_1, r_1 | z_2, r_2)$ describes the probability for finding the increment z_1 on scale r_1 provided that the increment z_2 is given on scale r_2. An important simplification arises if

$$p(z_1, r_1 | z_2, r_2; ...; z_n, r_n) = p(z_1, r_1 | z_2, r_2) \qquad \text{where} \qquad r_1 < r_2 < ... < r_n. \quad (4.4)$$

This expression defines a *Markovian process* evolving from r_{i+1} to r_i. The Marlovian property implies that only the most recent conditioning is relevant to the present probability. In consequence, each $n-$joint probability density can be expressed as a product of n single conditional probability density functions

$$p(z_1, r_1; \ldots; z_n, r_n) = p(z_1, r_1 | z_2, r_2) \cdot \ldots \cdot p(z_{n-1}, r_{n-1} | z_n, r_n) \cdot p(z_n, r_n) \quad (4.5)$$

For any Markovian process, the conditional PDF satisfies a master equation. Expanding the distribution function into a Taylor series, the evolution equation can be written as

$$-r \frac{\partial}{\partial r} p(z_r, r | z_0, r_0) = \sum_{k=1}^{\infty} \left(-\frac{\partial}{\partial z_r} \right)^k D^{(k)}(z_r, r) p(z_r, r | z_0, r_0). \quad (4.6)$$

The Kramers-Moyal coefficients $D^{(k)}(z_r, r)$ are defined as the limit $\Delta r \to 0$ of the conditional moments $M^{(k)}(z_r, r, \Delta r)$:

$$D^{(k)}(z_r, r) = \lim_{\Delta r \to 0} M^{(k)}(z_r, r, \Delta r),$$

$$M^{(k)}(z_r, r, \Delta r) = \frac{r}{k! \Delta r} \int_{-\infty}^{\infty} (\tilde{z}_r - z_r)^k p(\tilde{z}_r, r - \Delta r | z_r, r) d\tilde{z}_r \quad (4.7)$$

And can be calculated from the joint probability density functions, which are obtained from the experimental data by counting the number $N(\tilde{z}, z)$ of occurrences of the two increments $\tilde{z}_r$ and z_r. Assuming the error of $N(\tilde{z}_r, z_r)$ to be given by $\sqrt{N(\tilde{z}_r, z_r)}$, errors for the coefficients $M^{(k)}(z_r, r, \Delta r)$ can be derived.

A second level of simplification can be obtained if the noise included in the process is Gaussian distributed. In this case the coefficient $D^{(4)}(z_r, r)$ is zero and the equation (4.6) reduces to the Fokker-Planck equation

$$-r \frac{\partial}{\partial r} p(z_r, r | z_0, r_0) = \left\{ -\frac{\partial}{\partial z_r} D^{(1)}(z_r, r) + \frac{\partial^2}{\partial^2 z_r} D^{(2)}(z_r, r) \right\} p(z_r, r | z_0, r_0). \quad (4.8)$$

This equation describes the evolution of the conditional probability function form larger

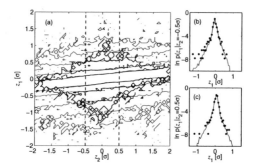

Figure 4.6: Test for Markovian properties of the methanol-electrolyte corresponding to $r_1 = 10\ \mu m$, $r_2 = 108\ \mu m$, $r_3 = 216\ \mu m$. (a) Contour lines of the single conditional PDFs $p(z_1, r_1 | z_2, r_2)$ (gray lines) and double conditional PDFs $p(z_1, r_1 | z_2, r_2; z_3 = 0, r_3)$ (black lines). (b) and (c) Cuts through the conditional PDF for $z_2 = \pm\sigma/2$. Lines: $p(z_1, r_1 | z_2, r_2)$, symbols : $p(z_1, r_1 | z_2, r_2; z_3 = 0, r_3)$.

to smaller length scales and thus also the complete $n-$scale statistics. The coefficient $D^{(1)}(z_r, r)$ is commonly denoted as the drift term, describing the deterministic part of the process, while the coefficient $D^{(2)}(z_r, r)$ is designated as the diffusion term, determined by the variance of a Gaussian, $\delta-$correlated noise.

4.4.1   Application to electropolished surfaces

In what follows we apply the method described above to the data series corresponding to surfaces electropolished with methanol-electrolyte [Haase *et al.*, 2004; Mora & Haase, 2004a,b; Mora *et al.*, 2004]. A preliminary analysis of multi-conditional PDFs suggests that the statistics of the height increments for scales larger than a certain threshold has Markovian properties. However, given the amount of data the verification of the condition (4.4) is only possible for $n = 3$. In Figure 4.6(a) the contour lines of the single conditional PDF $p(z_1, r_1 | z_2, r_2)$ (gray lines) are compared with the corresponding lines for the double conditional PDF $p(z_1, r_1 | z_2, r_2; z_3 = 0, r_3)$ for the increments $r_1 = 10\ \mu m$, $r_2 = 108\ \mu m$, $r_3 = 216\ \mu m$. The good correspondence of the contour lines gives preliminary evidence of the validity of the Markovian property (with $n = 3$). This correspondence can also be seen in the cuts through the conditional PDF for fixed $z_2 = \pm\sigma/2$ (Figures 4.6(b-c)). However, the Markovian property does not hold for any combination of increments r. For

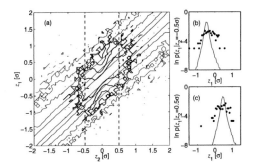

Figure 4.7: Test for Markovian properties of the methanol-electrolyte corresponding to $r_1 = 98\ \mu m$, $r_2 = 108\ \mu m$, $r_3 = 118\ \mu m$. a) Contour lines of the single conditional PDFs $p(z_1, r_1 | z_2, r_2)$ (gray lines) and double conditional PDFs $p(z_1, r_1 | z_2, r_2; z_3 = 0, r_3)$ (black lines). b) and c) Cuts through the conditional PDF for $z_2 = \pm\sigma/2$. Lines: $p(z_1, r_1 | z_2, r_2)$, symbols : $p(z_1, r_1 | z_2, r_2; z_3 = 0, r_3)$.

example for $r_1 = 96\ \mu m$, $r_2 = 108\ \mu m$, $r_3 = 118\ \mu m$, the Markovian test fails because the contour lines for the single and double conditional probabilities clearly does not coincide (see Figure 4.7).

Concerning the derivation of the Kramers-Moyal coefficients, Figures 4.8(a-b) shows the estimation $D^{(1)}(z_r, r = 108\ \mu m)$ and $D^{(2)}(z_r, r = 108\ \mu m)$ coefficients. Figures 4.9(c-d) shows the $D^{(3)}(z_r, r = 108\ \mu m)$ and $D^{(4)}(z_r, r = 108\ \mu m)$. This shows that $D^{(4)}$ is small and can be considered zero. Therefore this opens the possibility to express the evolution of the conditional probability by the Fokker-Planck equation (4.8).

4.5 Discussion

We presented numerical techniques based on wavelet analysis and stochastic methods for a characterization of complex structures resulting from electropolishing of brass surfaces. The multifractal scaling behavior was analyzed with the singularity spectra $D(h)$, which are estimated using the WTMM method. We have shown that it is possible to distinguish between the surface structures caused by two processes: roughening at small scales produced by the chemical dissolution of the metal and anisotropic structuring due to the

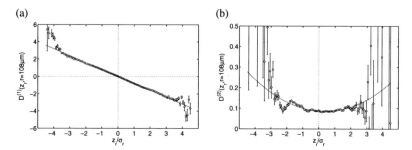

Figure 4.8: (a) Estimated values for the drift coefficient $D^{(1)}(z_r, r = 108)$ with uncertainty bars. The continuous curve represents a fit with a straight line $D^{(1)}(z_r, r = 108 \ \mu m) = -0.7948 z_r + 0.0053$. (b) Estimated values for the drift coefficient $D^{(2)}(z_r, r = 108)$ with uncertainty bars. The continuous curve represents a fit with a straight line $D^{(2)}(z_r, r = 108 \ \mu m) = 0.0081 z_r^2 - 0.0060 z_r + 0.0859$.

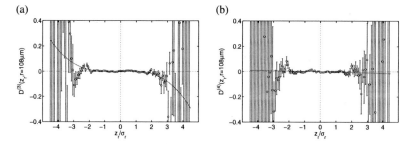

Figure 4.9: (a) Estimated values for the diffusion coefficient $D^{(3)}(z_r, r = 108)$ with uncertainty bars. The continuous curve represents a fit with a third order polynomial $D^{(3)}(z_r, r = 108 \ \mu m) = -0.0032 z_r^3 - 0.0012 z_r^2 + 0.0050 z_r + 0.0026$ (b) Estimated values for fourth Kramers-Moyal coefficient $D^{(4)}(z_r, r = 108)$ with uncertainty bars. The continuous curve represents a fit with a second order polynomial $D^{(4)}(z_r, r = 108 \ \mu m) = -0.00016 z_r^2 - 0.0029 z_r + 0.0005$.

combined action of the falling film and the gas bubbles.

For the electropolished surface corresponding to the methanol-electrolyte, it has been found evidence of Markov properties (within certain range of scales). This implies that the $r-$dependency of the height increments z_r can be regarded as a stochastic process evolving in r driven by deterministic and random forces. We have evaluated the Kramers-Moyal coefficients up to the fourth order directly from the data. Therefore a complete stochastic description of the surface can be given by a Fokker-Planck equation for the evolution of the conditional probability $p(z_r, r | z_0, r_0)$. Further work should include the verification of the procedure by means of the numerical solution of the Fokker-Planck equation.

Chapter 5

The temperature field in LJE

Because the laser-induced etching process is dominated by thermally activated chemical reactions, it is necessary to understand the main features of the temperature field. A detailed solution of the heat problem in the LJE experiment is not feasible due to its complexity. Instead, we use some simple considerations about the characteristic spatial and temporal scales of the heat diffusion in order to propose a "thermally thin" approximation for the temperature field. This approximation implies that for the conditions of the experiment, the heat transport generated by the laser beam can be treated like a two-dimensional problem. We estimate the temperature field depending on the laser power, and considering the depassivation temperature, we propose a procedure to account for the dependency of the etching front diameter on the power.

5.1   Preliminary considerations

The photons of the incident laser radiation are absorbed at the surface by interaction with electrons of the metal. These excited electrons collide with lattice phonons and with other electrons generating transfer of heat from the absorbing surface to the bulk material. This process of conversion from radiation energy to heat is considered to happen almost instantaneously. Furthermore, the thickness of the layer where radiation is absorbed is much smaller than the spatial scale of the temperature field, therefore light absorption can be considered as a surface process [Anisimov & Khokhlov, 1995; Ready, 1971].

The amount of absorbed light on each point depends on both the plane of polarization of the laser beam and the incidence angle. If the plane of polarization is parallel to the plane of incidence, the absorption has a maximum at certain angle, which is analogous to the Brewster's angle phenomenon that occurs for transparent materials. Whereas if the plane of polarization is perpendicular to the plane of incidence, the absorption always monotonically decreases with the incidence angle. The absorptivity for both polarizations approaches 0 as the angle of incidence approaches $\pi/2$ (see Figure 8.2(b)).

As examined in Chapter 2, the heat transfer appears in its three forms simultaneously: conduction, convection and radiation. Since the etching reaction is exothermic, it has to be considered as an additional heat source. The area that becomes hot is much broader than the laser spot, and the heat loss from the surface through convection and radiation may approach the same order of magnitude as the heat produced by laser light absorption [Ready, 1971]. Two types of convective transport have to be considered. First, free convection is produced because the etchant in contact with the surface is also heated, creating buoyancy of the etchant and micro-stirring. Secondly, forced convection is produced by the impinging etchant jet, which is difficult to quantify due to the lack of information about the hydrodynamics during the structuring. Concerning the radiative losses, a term proportional to the fourth power of temperature has to be considered according to the Stefan's law.

The values of the thermo-chemical constants for the materials are not known precisely and vary with the temperature. Therefore, the heat transfer problem is highly non-linear and becomes very difficult to solve, even numerically [Ready, 1971]. Solving such heat transfer equations with free boundary conditions changing in a complex way, as in the case of unstable kerfs, is out of the scope of this work.

5.2   Characteristic length and times

Dwell time is the characteristic time to be considered in the heat transport problem for LJE. Taking into account the ranges of feed velocities for the two experimental series analyzed through this work, $v_f \approx 6~\mu$m/s is a proper choice for the average feed velocity. From the average width of the experimental kerfs, we select the value $d_{front} = 30~\mu$m as the average value for the diameter of the etching front. Therefore, the characteristic dwell time is:

$$\bar{t}_{\text{dwell}} = \frac{\bar{d}_{\text{front}}}{\bar{v}_f} \approx 2 \text{ s.} \tag{5.1}$$

Certainly, two seconds is a lot of time for the heat diffusion in stainless steel. The thermal diffusivity κ measures how fast heat travels through the material. It is related to thermal conductivity K, specific heat C_P, and density ρ by:

$$\kappa = \frac{K}{\rho C_P}. \tag{5.2}$$

Through this chapter, we use for the stainless steel the following values: $\kappa = 3.77 \times 10^{-6}$ m^{2}s^{-1}, and $K = 16.3$ Wm^{-1}C^{-1}, which can found in the standard handbooks. These thermal constants are determined for conditions that can be significantly different to those found in the LJE experiment. Therefore, the computations presented here should agree only on the order of magnitude of the experimental ones. Heat diffuses away from an illuminated region of a material in a characteristic diffusion length Webb & Jones [2004]:

$$L_{\text{diff}} = 2\sqrt{\kappa\tau}, \tag{5.3}$$

where τ is the duration of the illumination. Using the estimated dwell time and thermal diffusivity of the stainless steel, the heat diffusion length is $L_{\text{diff}} \approx 5.5$ mm. This quantity is two orders of magnitude larger than the average width or depth of the obtained kerfs. In what follows, we analyze the case of the heating of stainless steel flat surface by a continuous Gaussian laser beam *without* etching. The absorbed laser Gaussian intensity as a function of radial position r is given by:

$$F(r) = F_0 \exp\left(-r^2/g^2\right), \tag{5.4}$$

where F_0 is the absorbed intensity at the center of the spot measured in W/m^2, and g is the *Gaussian radius* defined as $g = \sqrt{2}\sigma$, where σ is the standard deviation of the Gaussian. The temporal evolution of the temperature at the center of the laser irradiated

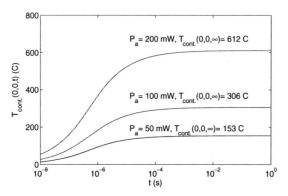

Figure 5.1: Temporal evolution of the temperature at the center of a Gaussian laser spot $T_{cont.}(0, 0, t)$, for three absorbed laser powers P_a. $T_{cont.}(0, 0, \infty)$ is the steady state temperature for each power P_a. The laser beam has the same diameter used in the LJE experiment.

spot ($x = 0$, $y = 0$) is given by (After [Ready, 1971], Chapter 3):

$$T_{\text{cont.}}(0, 0, t) = \frac{F_0 g}{K \sqrt{\pi}} \tan^{-1} \left(\frac{4\kappa t}{g^2} \right)^{1/2}, \tag{5.5}$$

where t is the time. Integrating the previous expression from $t = 0$ to $t = \infty$, the steady state value of the temperature is obtained:

$$T_{\text{cont.}}(0, 0, \infty) = F_0 g \sqrt{\pi}/2K. \tag{5.6}$$

In order to compare with the spatial scales of the LJE experiment, we can select the Gaussian radius $g = 2.83 \ \mu m$ to reproduce the diameter of the laser beam of the experiment ($d_{\text{exp}} = 4\sigma = 8 \ \mu m$). For example, if the absorbed power is 100 mW, the steady-state temperature is $T_{cont.}(0, 0, \infty) = 306$ C, which is small compared to the melting point for stainless steel (1400-1455 C). Figure 5.1 shows the temporal evolution of the temperature for the center of the laser spot computed using Equation 5.5 for three different absorbed laser powers P_a. The curves show how the steady state temperature $T_{cont.}(0, 0, \infty)$ is reached for times much shorter than the mentioned typical dwell times.

In conclusion, the characteristic diffusion length is much larger than the dimensions of the kerfs and the heating occurs at velocities much larger than the typical feed velocities.

Therefore, the local temperature profile generated by the moving laser beam is identical to that for static laser radiation without scanning, and different feed velocities lead only to changes in the beam dwell time, i.e. the temperature duration.

5.3 Thermally thin layer approximation

Given the characteristic times and lengths of the heat diffusion, the distribution of heat relevant to our problem can be considered almost two-dimensional. This "thermally thin" approximation [Ready, 1971] is justified because during the dwell time of the laser, the back surface of the sheet reaches approximately the same temperature as the front surface on which the radiation is incident. The approximation holds if $D^2/(4\kappa t)$ is much less than one, where D is the thickness of the film and t the time scale of interest. In the LJE experiment, the thickness of the sheet is $D = 200\ \mu m$ and the dwell time is of the order of 2 seconds. Therefore, $D^2/(4\kappa t) = 5.3 \times 10^{-4} \ll 1$ and the approximation is reasonable.

Within this approximation, the temperature field produced by the laser with constant absorbed intensity F_0 at the center of the Gaussian profile is given by:

$$T_{\text{cont.Gauss.}}(r, t, \bar{k}^2) = \frac{F_0 \kappa g^2}{KD} \int_0^t \frac{dt'}{4\kappa t' + g^2} \exp\left[-\kappa \bar{k}^2 t' - \frac{r^2}{4\kappa t'} + \frac{r^2 g^2}{4\kappa t'(g^2 + 4\kappa t')}\right],$$
(5.7)

where r is the radial distance from the center of the laser beam. For the derivation of this expression, losses by convection have been included by means of a term that gives rise to a linear heat transfer law across the surface. This is accounted by the parameter $\bar{k}^2 = 2H/KD$, where the surface conductance of the material H specifies the proportionality between the heat flow across the boundary and the temperature difference across the boundary. The value of H for heat transfer by free convection is a function of the geometry, the orientation of the surface, and the temperature difference. An approximate order of magnitude for plane surfaces in air is about $5 \times 10^{-8}\ \text{Wm}^{-2}\text{C}^{-1}$. The equation 5.7 may be reduced to dimensionless form by the substitutions:

$$\xi = r/g, \quad \varrho = \bar{k}^2 g^2/4, \quad \tau = 4\kappa t/g^2, \quad \tau' = 4\kappa t'/g^2,$$
(5.8)

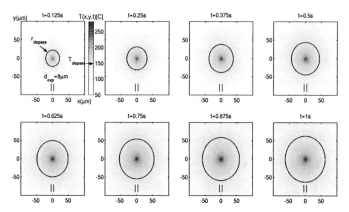

Figure 5.2: Temporal evolution of the surface temperature produced by a continuous Gaussian beam using a "thermally thin" layer approximation obtained by integrating the Equation 5.10. The growing circle is an isotherm at $T_{\text{depass}} = 150$ C. The two small and parallel lines are separated 8 μm, which is the laser spot diameter d_{exp}.

then

$$T_{\text{cont.Gauss.}}(r, t, \bar{k}^2) = \frac{F_0 g^2}{4KD} \theta(\xi, \tau, \varrho) =, \tag{5.9}$$

to obtain

$$\theta(\xi, \tau, \varrho) = \int_0^\tau \frac{d\tau'}{\tau' + 1} \exp\left[-\varrho\tau' - \xi^2/(\tau' + 1)\right]. \tag{5.10}$$

This equation can be used to obtain the temperature θ as a function of the generalized coordinates 5.8. Then, using Equation 5.9 and the specific values for the LJE experiments, an estimation of the temperature field can be done as it is shown in Figure 5.2. As in the previous section, the Gaussian radius $g = 2.83$ μm reproduces the diameter of the laser beam of the experiment. Each frame represents the top view of the temperature field where the temperature value is represented in gray scale. The circle is an isotherm at 150 C, which increases its diameter as the surface is heated. If the depassivation in the LJE experiment would occur at $T_{\text{depass}} = 150$ C, this growing circle could illustrate how the etching front is formed at early times. The diameter of the experimental Gaussian beam $d_{\text{exp}} = 8$ μm is indicated by the separation between the two small parallel lines

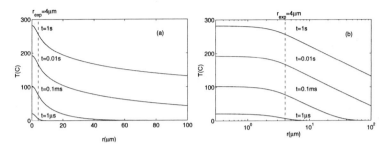

Figure 5.3: Temporal evolution of the surface temperature T as a function of the distance r from the center of a Gaussian beam using a "thermally thin" layer approximation. **(a)** Plot of T vs. r. The slashed vertical line is located at the Gaussian beam radius $g = 2.83\ \mu m$. The zone where the laser light is absorbed is not much bigger than $2g$. **(b)** Plot of T vs. $\log(r)$ reveals that beyond the zone where laser radiation is absorbed, the temperature is $T \propto -\log(r)$. The parameters are the same used in Figure 5.2.

that appear below the center in each frame. Note that for times much shorter than $t = 1$s, the heating process is fast and the zone affected by the heating is much larger than the diameter of the laser beam (for the arbitrarily chosen T_{depass}).

Figure 5.3(a) shows the profile of the same temperature field as a function of the radial distance r. The temperature within the laser spot increases very fast (the radius of the laser beam $r_{\text{exp}} = d_{mathrmexp}/2 = 4\mu$m is indicated by the vertical slashed line). Whereas for regions far away from the origin, the heating process is slower and the temperature is much lower. Figure 5.3(b) shows the same curves represented with the horizontal axis in a logarithmic scale, indicating that for times of the order of the average dwell time and regions beyond the Gaussian beam, the temperature is:

$$T(r) \propto -\log(r), \quad \text{for } r > r_{exp}, \text{ and } t \gtrsim 1\text{s}, \tag{5.11}$$

Figure 5.4 shows how to use the temperature profile to analyze the dependency of the size of the etching front with the laser power. Due to the fact that the temperature field is proportional to the power (through the parameter F_0 of the expression 5.7), it is possible to use one temperature profile at certain power to compute the rest of them. Therefore, the

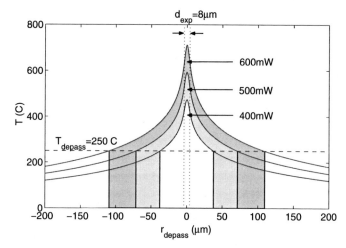

Figure 5.4: The curves show a typical profile of the temperature field produced by a continuous Gaussian laser for increasing laser powers (using the thin layer approximation). The slashed horizontal line defines the depassivation temperature above which the etching process can occur (in this figure arbitrarily defined to be T_{depass} =250 C). Within this scheme, the widths of the shaded areas determine the diameter of the corresponding etching front. The two vertical dotted lines indicate the diameter of the laser beam.

temperature profile for 1 W laser power has been scaled to obtain the temperature profiles for laser powers 400 mW, 500 mW, and 600 mW. The horizontal slashed line corresponds to an arbitrarily chosen depassivation temperature T_{depass} =250 C. Because the etching should occur only above T_{depass}, the limits of the etching front can be determined by finding where the temperature curve intersects $T_{\text{depass}} = 250$ C. As it is suggested in the figure by the three shaded areas, the diameter of the etching front increases with the power and is larger than the diameter of the laser spot as in the experiment. In Chapter 8 we use this scheme to simulate, in a probabilistic sense, the spatial distribution of the etching rate.

5.4 Discussion

The thermally thin approximation is a coarse description of the complex heat phenomena present in the LJE experiment. We considered only the heating of a stainless steel flat surface by a Gaussian laser beam and neglected the etching. We also excluded the exothermic heat produced by etching reactions and probably underestimated the convective effect of the etchant jet. The values of the thermal constants used in the computations should correspond within an order of magnitude to the actual ones.

In spite of all these limitations, we argue that this approximation provides a meaningful description of the metallic sample heating and formation of the etching front. Instead of formulating a detailed heat transfer model with all possible influences, we have made use of the fact that the temperature of the material is approximately uniform through its thickness in order to estimate the temperature field. Given the characteristic heat diffusion length and dwell times, we have found that this temperature field quickly reaches its steady state and can be considered independent of the feed velocity. Based on the estimated temperature profiles, we propose a scheme in which the power and depassivation temperature determine the diameter of the etching front.

Chapter 6

Universality and pattern formation in LJE

An important aspect of pattern-formation phenomena is that many of them have a universal character. This universality among diverse systems opens the possibility to describe them with similar phenomenological equations. Here we review the Kuramoto-Sivashinsky equation and how it has been modified to describe pattern formation features found in ion beam sputtering and water jet cutting. Linear stability analysis can be used to study the initial deviation from uniformity, and thus estimate the length scales of the resulting patterns.

Techniques of localized structuring of solid materials like abrasive water jet cutting, laser melting, and laser-induced jet-chemical etching (LJE) have in common that the resulting kerfs present periodic structures in the form of ripples or striations. Assuming universality, in Section 6.4 we use an experimental fact found in water jet cutting to propose a working hypothesis about the lengths of the ripples of a LJE experiment where the laser power is increased.

6.1 The Kuramoto-Sivashinsky equation

The Kuramoto-Sivashinsky (KS) equation, originally proposed by Kuramoto to describe the evolution of chemical reaction-diffusion systems [Kuramoto, 1984; Kuramoto & Tsuzuki, 1976] and by Sivashinsky in the study of flame fronts [Sivashinsky, 1979], has become a paradigm for pattern formation in extended systems. The KS equation provides the continuum description of interfaces appearing in diverse systems in which a periodic

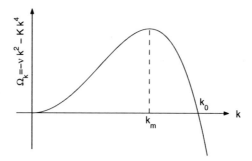

Figure 6.1: Stability of Fourier modes for the noisy KS equation. The modes from 0 to k_0 are unstable, while for $k > k_0$ the modes are stable. k_m corresponds to the mode that grows faster and defines the dominant ripple length.

pattern develops with a preferred wavelength, which afterwards evolves into a disordered state. The *noisy* version of the KS equation for a 1+1 dimensional interface characterized by the height $h(x, t)$ is given by

$$\frac{\partial h(x, t)}{\partial t} = \nu \nabla^2 h(x, t) - K \nabla^4 h(x, t) + \frac{\lambda}{2} [\nabla h(x, t)]^2 + \eta(x, t). \tag{6.1}$$

The surface tension coefficient ν is negative, which opens the posibility that the system is linearly unstable. K is a positive coefficient associated with the surface diffusion. The strength of the non-linearity is given by λ, which couples the stable and unstable modes, thus stabilizing the system [Rost & Krug, 1995]. In the limit of long time and long scales, the system leads to a rough morphology. The term $\eta(x, t)$ is a Gaussian white noise with zero mean $\langle \eta(x, t) \rangle = 0$, and short-range correlations described by

$$\langle \eta(x, t) \eta(x', t') \rangle = 2D \delta(x - x') \delta(t - t'), \tag{6.2}$$

where D is a constant. The competition between the unstable erosion term $\nu \nabla^2 h$ and the smoothing term $-K \nabla^4 h$ leads to the existence of a finite band of linearly unstable Fourier modes in the surface. Performing a linear stability analysis on Equation (6.1), one finds that the amplitude for solutions of the form

$$h_0(x, t) \sim \exp(\Omega_k t) \sin(kx) \tag{6.3}$$

is characterized by the rate

$$\Omega_k = -(\nu k^2 + K k^4). \tag{6.4}$$

This expression is plotted in Figure 6.1. The unstable modes correspond to the positive values of Ω_k with wave numbers between 0 and k_0, where

$$k_0 = \sqrt{|\nu|/K}. \tag{6.5}$$

The modes for $k > k_0$ are stable. The most unstable mode is the one whose amplitude exponentially dominates all the others, and corresponds to the value of k, at which Ω_k reaches the highest positive value

$$k_m = \sqrt{\frac{|\nu|}{2K}}, \tag{6.6}$$

which corresponds to a wavelength of

$$l_m = 2\pi/k_m = 2\pi\sqrt{2K/|\nu|}. \tag{6.7}$$

This is the initial characteristic length expected in systems described by Equation (6.1) or modifications of it. In the next section we present two of such systems. The KS equation produces at large times a turbulent, chaotic steady state, in which the instabilities are transfered by the non-linear term to a larger wavenumber k, where they are dissipated by the fourth-order derivative [Halping-Healy & Zhang, 1995]. It has been debated [Tok *et al.*, 2004] whether or not the asymptotic scaling properties of the KS equation in 2D+1 dimensions are related with the Kardar-Parisi-Zhang (KPZ) equation [Kardar *et al.*, 1986], which has been used intensively in kinetic roughening models [Barabási & Stanley, 1995].

6.2   Ion beam sputtering

Ion beam sputtering is the removal of material from the surface of solids through the impact of energetic ions. Among other applications, sputter erosion is routinely used for etching patterns in integrated circuits and device packaging. The evolution of solid surface topography is governed by the competition between the dynamics of surface roughening due to erosion and smoothing due to surface diffusion (see general reviews in [Carter,

2001; Valbusa *et al.*, 2002]). As a consequence of this process, the surface tends to create almost periodic ripples aligned parallel or perpendicular to the bombarding ion beam. For a complete discussion of ion sputtering in crystalline and amorphous materials see [Makeev & Barabási, 2004a; Makeev *et al.*, 2002], [Makeev & Barabási, 2004b]. For pattern formation in semiconductor surfaces by ion sputtering under normal incidence see [Facsko *et al.*, 2002, 2004].

According to Sigmund [Sigmund, 1969], the incoming ion penetrates the surface and delivers its kinetic energy by means of a cascade of collisions within a Gaussian spatial distribution centered about a finite distance from the point of initial impact. For geometrical reasons the absorbed energy concentrates more in regions of the surface with positive curvature, which means that erosion at valleys is favored over erosion at peaks. Bradley and Harper (BH) found that combining this curvature-dependent erosion with the surface smoothing mechanism due to surface diffusion, a linear evolution equation for a two dimensional interface $h(x, y, t)$ can be derived [Bradley & Harper, 1988]:

$$\frac{\partial h(x, y, t)}{\partial t} = -v(\theta) + \nu_x(\theta)\partial_x^2 h(x, y, t) + \nu_y(\theta)\partial_y^2 h(x, y, t) - K\nabla^4 h(x, y, t), \quad (6.8)$$

here $\nu_x(\theta)$ and $\nu_y(\theta)$ are the effective surface tensions generated by the erosion process, dependent on the angle of incidence of the ions θ; K is the relaxation rate due to surface diffusion; and $v(\theta)$ is the interface velocity. The competition between the erosion and the surface mechanisms can produce ripples, which according to linear stability analysis have a wavelength $l_j = 2\pi\sqrt{2K/|\nu_j|}$, where j refers to the direction (x or y) along which the associated $|\nu_j|$ is the largest [Valbusa *et al.*, 2002].

The BH theory predicts the ripple wavelength and orientation, but it does not account for the stabilization of the ripples and subsequent kinetic roughening. To solve this situation, it has been shown that a two dimensional version of the noisy Kuramoto-Sivashinsky equation (6.1) contains the lowest order terms modeling the microscopic mechanisms of the ion sputtering [Cuerno & Barabási, 1995]. In particular, the non-linear term $\lambda/2(\nabla h)^2$ is responsible for limiting the exponential growth of the linear instability. Currently various anisotropic versions of the KS equation and inclusion of high-order terms have been proposed to explain experimental deviations from the established theory (see [Castro *et al.*, 2005] and references therein).

6.3   Water jet cutting

Abrasive water jet cutting is a technique used to cut hard materials. A focused water jet of diameter $\approx 1mm$ containing abrasive particles like sand or garnet, is directed at high velocities against the samples [Momber & Kovacevic, 1998]. The cut is performed moving the jet with a constant feed velocity. For relative high feed velocities, the walls of the cut present unwanted periodic striations, which are found to be of the same order of magnitude as the jet diameter. With low feed velocities, the striations are reduced but the operational costs increase. Similar striation phenomena are also observed in laser-driven melting [Schulz *et al.*, 1999], or gas-assisted laser cutting [Tirumala Rao & Nath, 2002].

Friedrich *et al.* proposed a phenomenological model based on a modified version of the KS equation in order to explain the occurrence of striation patterns in water jet cutting [Friedrich *et al.*, 2000b; Radons *et al.*, 2004]. In a frame moving with the water jet, such equation reads

$$\frac{\partial h(x,t)}{\partial t} + u\nabla h(x,t) = v(x)\left(\frac{1}{1+[\nabla h(x,t)]^2} + \alpha\nabla^2 h(x,t) + \beta\nabla^4 h(x,t)\right), \quad (6.9)$$

where the second term of the left hand side of the equation is a convective term related to the feed velocity u; $v(x)$ is a Gaussian distribution that represents the action of the localized jet. The non-linear term $1/(1+[\nabla h(x,t)]^2)$ it has been found to represent the angle dependency of the removal rate for brittle materials [Finnie & Kabil, 1965]. The non-linearilties provide the mechanism to control eventual instabilities. The terms related to the variables α and β can be considered analogous to the terms related to the variables ν and K of Equation (6.1). In the case of an infinitely extended beam $v(x,t) = $ const (homogeneous case), an instability develops when $\alpha < 0$ and $\beta < 0$. Applying linear stability analysis, the fastest growing Fourier mode has a wavelength

$$l_m = 2\pi/k_m = 2\pi\sqrt{2\beta/\alpha}. \tag{6.10}$$

For the inhomogeneous case, the localized jet is taken as $v(x) = N\exp[-x^2/(2\sigma^2)]$ with σ the standard deviation, and N measuring the effective removal rate. If the effective diameter of the jet 4σ is chosen to be of the same order as l_m, a numerical integration of Equation (6.9) shows for large feed velocities, stationary fronts; while for lower feed velocities, ripples with lengths close to l_m are obtained [Friedrich *et al.*, 2000b; Radons *et al.*, 2004].

(a)   (b)

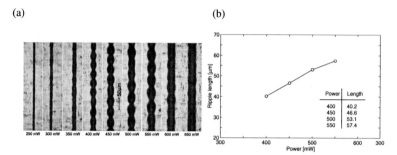

Figure 6.2: (a) Optical microscope images showing a series of kerfs structured by LJE with powers from 250 to 650 mW (etchant 5M H_3PO_4, feed velocity 6 μm/s, etchant jet velocity 190 cm/s). The kerfs for 250, 300 and 350 mW are the result of a uniform etching front that widens with power. For powers greater than 350 mW, an instability in the etching front appears producing rippled kerfs (b) Ripple length of the kerfs from 400 to 550 mW ($\circ$ symbols). (Images courtesy Thomas Rabbow, IAPC-Bremen).

6.4   A hypothesis about intrinsic ripple formation in LJE

In this section we use a universality assumption in a different sense to analyze a LJE experiment where a ripple regime appears when the laser power is increased. Since the microscopic mechanism that produces the ripples is currently unknown, we draw analogies from other pattern formation systems in order to propose a simple working hypothesis, which describes the ripple length dependence with the laser power.

Figure 6.2 shows a series of experiments performed at constant feed velocity $v_f = 6$ μm/s. For laser powers 250, 300, and 350 mW an approximately stationary etching front produces uniform kerfs with increasing width and depth (see the first three kerfs in Fig. 6.2). Certainly, more delivered power means higher temperatures, and although the laser spot diameter is always the same, the etching front become wider. For powers larger than 350 mW, an instability appears producing ripples with increasing wavelength (and width). For powers larger than 600 mW the ripples seem to overlap as the kerfs become deeper. The ripples look like a product of thermal runaways that reach broader areas than in the stationary etching front case.

It is necessary to determine whether the ripples are of intrinsic nature or result of periodic external influences. As shown in Chapter 3, there are two external sources that produce periodic structures. First, the peristaltic pump introduces vibrations in the etchant jet in the range of 1-3 Hz that can be detected in the power spectra of the electrochemical potential time series, which corresponds to periodic surface structures smaller than 10 μm. Second, the xy-stage introduces periodic structures of 40 μm. Nevertheless, the set of wavelengths of the examined ripples (see the table in Figure 6.2(b)) is not related with the wavelengths that would correspond to structures caused by either the etchant pump or the controlling mechanism of the xy-stage. In consequence, these causes can be discarded as external triggers of the obtained ripples.

From the point of view of the theory of pattern formation in continuous media, the appearance of ripples originating from instabilities is not unexpected because the etching front is formed in a system far from equilibrium due to the continuous and combined action of the laser beam and etchant jet. In general, one class of mechanisms for such instabilities arises from the existence of constraints and conservation laws in the system [Cross & Hohenberg, 1993; Walgraef, 1997]. A typical example of this is the Rayleigh-Bérnard convection of a fluid placed between flat horizontal plates separated by a distance d. The fluid is driven by maintaining the low plate at a temperature slightly above the upper plate temperature, creating a gradient of temperature ΔT. The mass of heated fluid moves upwards by buoyancy and the upper plate imposes a constraint on this movement. Above a threshold of the temperature difference ΔT, an instability produces a pattern in which the fluid raises in some regions and falls in others, creating rolls with a characteristic horizontal length scale close to d (the separation of the plates).

Another example of a spatial constraint that determines the scale of the resulting pattern can be seen in the case of water jet cutting, where the length of the striation patterns is of the *order* of the water-jet diameter (see [Momber & Kovacevic, 1998; Radons *et al.*, 2004]). This fact has been used as a criterion to distinguish between ripples triggered externally from intrinsic ones. In the LJE experiment of Figure 6.2, the average diameter of the etching front (either uniform or unstable) increases with the laser power due to the spread of more heat. We formulate our *working hypothesis* as follows: when there are conditions for instabilities (which occur only after a laser power threshold), *the length of the resulting ripples is proportional to the average diameter of the etching front*. This agrees with the fact that the ripple length increases with the power.

An analogy to support this hypothesis can be drawn from the Rayleigh-Bernard convection. The plates distance d imposes a spatial constraint that defines the horizontal size of the convection cells. In the LJE case, the average diameter of the etching front would be the spatial constraint that determines the length of the ripples. A closer analogy can be made with the water jet case: the characteristic length of the striation pattern is proportional to the jet diameter [Radons *et al.*, 2004]. In the LJE case the diameter of etching front depends proportionally on the power, thus the resulting ripple length increases with power.

6.5 Discussion

The Kuramoto-Sivashinsky equation is one of the simplest one-dimensional PDE's that exhibits complex dynamical behavior. We have reviewed two different physical systems that can be treated with modifications of the Kuramoto-Sivashinsky equation: ion beam sputtering and water jet cutting. Both descriptions have in common the appearance of a ripple regime that eventually is limited by the effect of the non-linear terms. Using linear stability analysis, an estimation of the length of the initial ripples can be expressed in terms of the parameters of the corresponding modified KS equation. Concerning the ion beam sputtering experiments, it has been established that the origin of the instability is the curvature dependency of the erosion processes. In the forthcoming chapters, assuming universality, we adapt a discrete model initially proposed for ion beam sputtering for the LJE experiment.

Analyzing the qualitative behavior of a series of LJE experiments with increasing laser power, we have claimed that the resulting ripples are of intrinsic nature. Based on analogies with other systems of pattern formation, we have proposed the hypothesis that the ripple length is proportional to the average diameter of the etching front (which is proportional to the power). The simulations presented later in Chapter 8 are found to be compatible with this assumption.

Chapter 7

The extended model

Here we present the *extended model* [Mora *et al.*, 2005a,b], which is a modification of a discrete model proposed by Cuerno, Makse, Tommassone, Harrington and Stanley (CMTHS) for ripple formation in ion beam sputtering [Cuerno *et al.*, 1995]. In the next chapter, assuming universality we use the extended model to simulate unstable behavior in the etching process of the LJE experiment. There, we will combine the extended model and the thermally thin layer approximation presented in Chapter 5 into the *LJE simulation* in order to represent the temperature-dependent localized etching of the corresponding process.

The CMTHS model is a stochastic 1D discrete solid-on-solid (SOS) model based on the competition between erosion and surface diffusion processes. The model reproduce the typical evolution of the sputtered surfaces: a ripple regime that appears at early stages, followed by a regime where the surface is roughened. Here we show why the features of the ripple regime of the CMTHS model are not suitable to be used it in the LJE simulation. We propose an improvement of the algorithm that simulates the curvature-dependent erosion process, obtaining a ripple regime characterized by longer and smoother ripples. This is what we call the *extended* model.

In this chapter we start with a brief review of solid-on-solid models, the kinetic Monte-Carlo method, and the scaling theory of far from thermal equilibrium interfaces, which are the theoretical basis for the CMTHS and extended models. Then, we analyze the scaling behavior and the evolution of the interface for both models. Finally, we present a procedure based on the master equation for establishing a continuous equation corresponding to the dynamics of the extended model. This equation turns out to be a modified noisy Kuramoto-Sivashinsky equation.

7.1   Solid-on-solid models and the kinetic Monte Carlo method

All the macroscopic features of any structured surface emerge from interactions at the atomic level. A complete quantum mechanical description at such a scale can be accomplished by the *density functional theory* (DTF) [Biehl, 2004; Kratzer & Scheffler, 2001]. In order to describe the collective behavior of thousands of atoms, *molecular dynamics* (MD) techniques account for the Newtonian interaction of all particles under potentials that can be inferred by DTF or be adjusted from experimental data. The main shortcoming of the MD simulations is that the addressed temporal scales are less than 10^{-6} s. In consequence, thermally activated processes that occur on larger temporal scales cannot be simulated with MD. Diverse Monte Carlo techniques have been developed to cover these temporal scales and can account for the behavior of millions of atoms [Landau & Binder, 2002; Newman & Barkema, 1999].

An important branch of the study of interfaces outside thermal equilibrium deals with the formulation of lattice discrete *solid-on-solid* models (SOS), which are well suited for systems that are thermally activated. They were proposed initially like a toy model for explore the statistical properties of system of many body systems, of for simulate the growth of crystal surfaces based on transition rates. The SOS can be considered as a restricted version of lattice gases in which every site is either vacant or occupied by a single atom, and simple interactions with another neighboring atoms can represent the dynamics of the system [Burton *et al.*, 1951; Newman & Barkema, 1999; Weeks, 1980]. Imposing the condition that every occupied site is directly on top of another occupied site, overhangs are prohibited. The SOS models simplify both the simulations and analytical approaches, where descriptions with single valued functions are preferred. Furthermore, standard techniques for measuring surfaces topographies, e.g. the laser profilometry, can only give a unique value for the height at certain position. However, it is clear that in the real physical process, the surfaces could develop complex geometries with overhangs, and even porous structures, that are difficult to measure.

Although solid-on-solid lattice models with nearest neighbor interactions cannot describe any material faithfully, they serve as prototype systems for the investigation of many basic and qualitative features of the surface dynamics [Biehl, 2004]. Deposition and erosion processes can be represented by the creation and annihilation of atoms at random positions on the surface. Surface diffusion processes can be represented by hops between lattice sites. Within the *kinetic Monte Carlo* method such random events are simulated with probabilities according to the corresponding event rates. These rates are guessed through the use of all the available experimental and theoretical information [Metiu *et al.*, 1992]. By varying the rates and comparing the results to the data, it is possible to find the physical quantities that are essential for reproducing the experimental observations. An additional advantage is that the kinetic Monte Carlo method naturally reproduces the fluctuations resulting from the stochastic nature of the physical processes.

7.2 Scaling theory of rough surfaces

In recent years, the concepts of fractal geometry, scaling and universality have been integrated in order to analyze the roughening of surfaces far from thermodynamical equilibrium [Barabási & Stanley, 1995; Family & Vicsek, 1991; Halping-Healy & Zhang, 1995; Meakin, 1998; Ódor, 2004]. It has been found that the roughness of some of such surfaces follows simple scaling laws, and by means of the corresponding scaling coefficients, universality classes can be defined. Discrete models have been proposed for describe growth processes, such as the Eden process [Eden, 1961], ballistic deposition (BD) [Vold, 1959], and growth of various restricted solid-on-solid models [Kim & Kosterlitz, 1989; Meakin *et al.*, 1986] among others. Those models have been classified within the Edwards-Wilkinson (EW) [Edwards & Wilkinson, 1982], Kardar-Parisi-Zhang (KPZ) [Kardar *et al.*, 1986] and other universality classes.

If the surface is represented by the one-dimensional single valued height function $h(x, t)$, where $x = 1, \ldots, L$, its roughness can be described by the evolution of the *interface width* $W(L, t)$, defined by the standard deviation of the height $h(x, t)$:

$$W(L,t) = \sqrt{\frac{1}{L} \sum_{x=1}^{L} [h(x,t) - \overline{h}(t)]^2} \tag{7.1}$$

where L is the linear size of the sample, and $\overline{h}$ is the *mean surface height*:

$$\overline{h}(t) \equiv \frac{1}{L} \sum_{x=1}^{L} h(x,t) \tag{7.2}$$

Starting from a flat surface, the interface width initially grows, and after a crossover time $t_\times$ it gets saturated. Namely,

(i) At early times $t \ll t_\times$,

$$W(L,t) \sim t^\beta, \tag{7.3}$$

where β is called the *growth exponent*.

(ii) For $t \gg t_\times$, the interface width saturates in a value W_{sat} that depends on the system size,

$$W_{\text{sat}}(L) \sim L^\alpha \tag{7.4}$$

The coefficient α is called the *roughness* exponent. This saturation is due to a finite size effect (see [Barabási & Stanley, 1995]). The dependency of the crossover time $t_\times$ with the system size follows

$$t_\times \sim L^z, \tag{7.5}$$

where $z = \alpha/\beta$ is the *dynamic exponent* that characterizes the crossover from the growth regime to the steady state.

All the information about scaling expressed in the previous equations can be combined in the *Family-Vicsek scaling relation* [Family & Vicsek, 1985]:

$$W(L,t) \sim L^\alpha f(t/L^z) \tag{7.6}$$

The function $f(u)$ defines two different scaling regimes depending on its argument u. For small u, $f(u \ll 1) \sim u^\beta$, which implies (7.3). For $t \gg t_\times$ the width saturates and $f(u \gg 1) \sim \text{constant}$, which implies (7.4). Equation (7.6) allows to define *universality classes*. The values of the coefficients α and β are independent of many details of the system, thus they are universal in the sense that they can be shared by different systems.

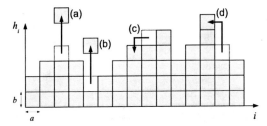

Figure 7.1: Diagram showing the lattice representation of a surface exposed to erosive and diffusive actions. Sites submitted to the erosion rule (a) and (b): The probability of being eroded is larger for the "valley" in (b) than for the "peak" in (a) according to an estimation of the curvature described in Sections 7.3.1 and 7.4.1. Diffusive movements (c) and (d): the probability of the surface diffusion in (c) is close to one, while for (d) it is very small according to a mechanism for creating a positive surface tension described in Section 7.3.2.

In contrast, other quantities like $t_\times$, or W_{sat} are non-universal because they depend on the details of the system. Recently, *anomalous roughening* has been found to occur in many growth models as well as experiments. In these cases the measurable roughness coefficient α_{loc} is different from α and must satisfy a different scaling law (see [Castro *et al.*, 1998; Krug, 1997; Ódor, 2004] and references therein).

7.3   The CMTHS model for ion sputtering

In this section we review the discrete CMTHS model for ion beam sputtering [Cuerno *et al.*, 1995]. We reproduce the scaling analysis of the interface width performed in the original work. Because we are interested in characteristics of the ripple regime, we examine the temporal evolution of the surface morphology with standard Fourier techniques. A close inspection of the curvature algorithm demostrates the necessity to propose a modification, which is the base of the extended model presented in the next section.

The material to be eroded is represented in 1+1D by a lattice composed of cells of width a and height b, and the surface is represented by the integer valued height $h_i(t)$, where $i = 1, \ldots, L$ (see Figure 7.1). The system size L is the number of cells in the horizontal direction. Since the surface is assumed to have no overhangs, all the sites below the sur-

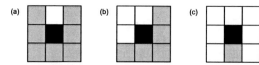

Figure 7.2: Three cases of the "box rule" used in the CMTHS model. The black cell is the candidate selected to be eroded and the gray boxes represent the occupied neighboring boxes. The probability of being eroded p_κ is proportional to number of gray boxes. Normalizing to 7, the box in (a) corresponds to $p_\kappa = 7/7 = 1$, (b) corresponds to $p_\kappa = 5/7$, and (c) corresponds to $p_\kappa = 1/7$. These three cases represent a "valley", a "flat part", and a "peak", respectively.

face are occupied with cells; whereas all the sites above are empty (the SOS condition). The temporal evolution of this virtual surface takes into account rules for representing two basic physical processes: the erosion triggered by the incoming ions carring kinetic energy and the thermally-activated surface diffusion processes. The algorithm starts with a flat surface (one of the possible initial conditions), selects a site, and applies with probability f the erosion rule and with probability $1 - f$ the surface diffusion rule. The process of selection of sites or rules to be applied is performed using a random number generator. For reproducing the CMTHS model results, and for our extension of the model, we implemented the random-number generation routines discussed by Press et al. (see Chapter 7 in [Press *et al.*, 1992]). In order to apply the rules to the sites $i = 1$ and $i = L$, periodic boundary conditions are established.

7.3.1 Erosion rule - CMTHS model

The erosion probability p_e for a cell at the site i is computed as the product $p_e = p_\kappa Y_i$. The quantity p_κ corresponds to the probability of being eroded depending on the curvature of the surface at the site, and accounts for the unstable erosion mechanism that exists in the physical system due to the finite penetration depth of the bombarding ions into the eroded substrate (see Section6.2). The term Y_i accounts for the efficiency of the sputtering yield (explained in detail later).

The value of p_κ is larger for positive curvatures than for negative ones (see Figure 7.1 (a) and (b)). In order to force a larger erosion rate in the "valleys" (regions with positive curvature) than in the "peaks" (regions with negative curvature), the algorithm uses a "box rule" to assign an erosion probability p_e for the selected site, counting the number of nearest neighbors around the site within a 3×3 box (see figure 7.2). This is equivalent to an estimation of the second derivative of the surface in the site i using the height values of the nearest neighbors (restricted to the box limits). This second derivative is a coarse approximation of the curvature and naturally presents limitations.

First of all, the box rule accounts for only 7 cases: $p_\kappa = 6/7$ and $p_\kappa = 7/7$ for positive curvature ("valleys"), $p_\kappa = 5/7$ for one zero curvature value ("flats"), and $p_\kappa = 1/7$, $2/7$, $3/7$ and $4/7$ for negative curvature ("peaks"). Secondly, because the rule does not account for height values of the neighboring sites that are outside of the 3×3 box, it cannot distinguish among many possible values of the curvature. This truncation of the value of curvatures is a source of non-linear behavior as discussed in Section 7.5. In Section 7.4.1 we propose a replacement for the estimation of the curvature and probability p_κ that can provide a wider spectrum of values and an additional control of the regimes and obtained ripple lengths.

The dependency of the erosion rate on the angle of incidence φ between the beam and a tilted portion of the surface is described by the sputtering yield function:

$$Y_i = Y(\varphi) = y_0 + y_1\varphi^2 + y_2\varphi^4, \tag{7.7}$$

where φ is the incidence angle formed by the incoming beam and the normal direction defined on the surface at the site i, and can be estimated with $\varphi = \arctan[(h_{i+1} - h_{i-1})/(2a)]$. The yield function $Y(\varphi)$ introduces a non-linear behavior, which becomes more relevant at the late regime of the evolution when large slopes develop, and in consequence the interface is roughened.

In the original article [Cuerno et al., 1995], the function $Y(\varphi) = 0.5 - 0.479\varphi^2 + 0.979\varphi^4$ has been selected to generically represent the sputtering yield of the experiments. In what follows, the model with this yield function is called the *full* CMTHS model. Whereas, the CMTHS model with the yield function $Y(\varphi) = 1$ neglects the sputtering yield dependency on the incidence angle and approximately holds at the early stages of the evolution.

7.3.2   Surface diffusion rule - CMTHS model

Surface diffusion is a relaxation mechanism always present at a nonzero temperature. Earlier models of surface diffusion in molecular-beam epitaxy (MBE) were proposed by Wolf and Villain [Wolf & Villain, 1990], and Das Sarma and Tamborenea [Das Sarma & Tamborenea, 1991], where a newly deposited particle is allowed to move to a nearest-neighbor site if this move results in a higher coordination number (number of neighbors). However, those models neglect the fact that all surface atoms should, in principle, be mobile. Therefore, two different solid-on-solid models of surface diffusion in MBE have been proposed: Arrhenius and Hamiltonian models (see review in [Yewande *et al.*, 2005]). In the Arrhenius models, the surface diffusion is a nearest-neighbor hopping process, with a hopping rate of an Arrhenius form (see [Smilauer *et al.*, 1993]). These models reproduce the Schwoebel effect, which is an additional potential barrier at the border of a terrace that repulses a diffusing particle from a down step, while the uphill diffusion current is preferred [Schwoebel, 1969].

The Hamiltonian models of surface diffusion are based on a thermodynamic interpretation of the diffusion process (see [Krug *et al.*, 1993] and [Siegert & Plischke, 1994]). A particular case is used in the CMTHS model, as follows. The surface energy is defined through the Hamiltonian:

$$\mathcal{H} = \frac{J}{b^2} \sum_{i=1}^{L} (h_i - h_{i+1})^2 \tag{7.8}$$

where J is a coupling constant through which the nearest neighbor sites interact, and b is the height of the cells. For each step, a site i and one neighbor site ($i - 1$ or $i + 1$) are randomly selected. The trial move is a cell hopping from the site i to the site $i \pm 1$. The surface energy $\mathcal{H}$ is computed before and after the hop, and the hop is allowed with the hopping probability:

$$w_i^{\pm} = \frac{1}{1 + \exp[\Delta\mathcal{H}_{i \to i\pm1}/(k_B T)]}, \tag{7.9}$$

where $\Delta\mathcal{H}_{i \to i\pm1}$ is the energy difference between the final and initial states of the move, k_B is the Boltzmann constant and T is the temperature. These transition rates are of the Glauber type [Glauber, 1963]. Diffusion movements that produce final states with lower surface energy are then highly preferred (this is illustrated in Figures 7.1(c) and (d)). In this surface diffusion model, no Schwoebel effect is present.

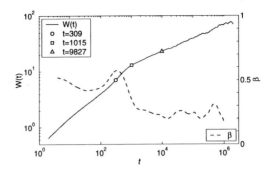

Figure 7.3: Temporal evolution of the interface width $W(t)$ (continuous line, scale on the left) for the full CMTHS model (yield function $Y(\varphi) = 0.5 + 0.979\varphi^2 - 0.479\varphi^4$). The slashed line (scale on the right) is the consecutive slope of the width, showing the value of the growth exponent β. Model parameters: $L = 2048$, $f = 0.5$, and $J/(k_BT) = 5$. The morphology of the surface at times corresponding to $t = 309$ (○), $t = 1015$ (□), and $t = 9827$ (△) is analyzed in Figure 7.4.

7.3.3 Scaling of the full CMTHS model

For the purpose of analyzing the scaling properties of the interface width $W(t)$ defined in Equation (7.1), we show in Figure 7.3 its temporal evolution (continuous line, logarithmic scale on the left). This curve is obtained from an ensemble average of over 30 different realizations of the random number generator. In order to determine regions of scaling behavior given by Equation (7.3), the growth exponent β is computed from $W(t)$ by a consecutive slopes method (see [Barabási & Stanley, 1995], page 305). The evolution of the growth exponent β is depicted with a slashed line (scale on the right). The probability of invoking the erosion rule is $f = 0.5$ (the surface diffusion rule is invoked with probability $1 - f = 0.5$), and the constant associated with the surface diffusion is $J/(k_BT) = 5$. For all the figures and analyses of this chapter, the time unit t is chosen to correspond to L erosion or surface diffusion rule invocations.

The growth exponent β varies permanently, but some interesting regions can be distinguished. At very early times $\beta \approx 0.5$, which is the characteristic value for random erosion. After $t \approx 100$, $W(t)$ scales with $\beta > 0.5$ due to instability originated by the curvature-dependent erosion rule. For $300 \lesssim t \lesssim 1000$ there is a fast decay of the value of the exponent β, which marks the onset of the non-linear effect of the sputtering yield.

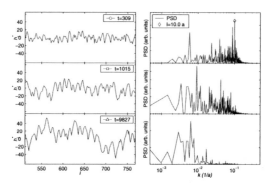

Figure 7.4: Three different stages of the evolution of the surface for the full CMTHS model ($Y(\varphi) = 0.5 + 0.979\varphi^2 - 0.479\varphi^4$) corresponding to the times referred in Figure 7.3 for a particular realization of the random number generator. Model parameters: $L = 2048$, $f = 0.5$, and $J/(k_BT) = 5$. The left column shows a 256 point-long portion of the height h_i^*, which is the height minus its average value at each time $h_i^* = h_i - \langle h_i \rangle$. The right column shows the power spectral density (PSD) computed with all the 2048 points of the surface versus the wave number k for each of the three stages. The PSD values are displayed in a linear scale.

For times $1000 \lesssim t \lesssim 10^6$ the exponent β fluctuates between 0.25 ± 0.05.

Figure 7.4 shows a portion of 256 points of the surface for the times $t = 309$, $t = 1015$, and $t = 9827$ indicated in Figure 7.3. The right column shows the power spectral density (PSD) taking into account all the $L = 2048$ cells of the surface. For $t = 309$, a noisy interface presents some irregular ripples. According to the power spectral density (PSD), these ripples have an average length $l \approx 10a$ (indicated by the $\diamond$ symbol). For $t = 1015$ the ripples are more noticeable, but they are irregular and they are modulated by small wave numbers as shown in the PSD. For $t = 9827$ the ripples are irregular and strongly modulated by small wave numbers. For later times, the surface is well inside of the non-linear regime and it can be considered rough.

For this full CMTHS model, we systematically varied the parameters f and J/k_BT in order to find more regular ripples with a length larger than $l = 10a$, but the obtained lengths were not long enough to simulate the ripples appearing in LJE in a meaningful way. This is the reason for proposing the extension of the CMTHS model discussed later.

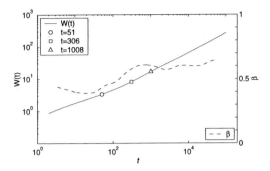

Figure 7.5: Temporal evolution of the interface width $W(t)$ (continuous line, scale on the left) for the CMTHS model with $Y(\varphi) = 1$. Model parameters: $L = 2048$, $f = 0.5$ and $J/(k_B T) = 5$. The slashed line (scale on the right) is the consecutive slope of the width, showing the value of the growth exponent β. Model parameters: $L = 2048$, $f = 0.5$ and $J/(k_B T) = 5$. The morphology of the surface at times corresponding to $t = 51 (\circ)$, $t = 306$ ($\square$), $t = 1008$ ($\triangle$) is analyzed in Figure 7.6.

7.3.4 Scaling of the CMTHS model with $Y(\varphi) = 1$

In order to study the ripple regime without the influence of the non-linearities originated by the sputtering yield function (7.7), we analyze the evolution of the surface with $Y(\varphi) = 1$, keeping all the other model parameters equal. Figure 7.5 shows the evolution of $W(t)$ (continuous line, scale on the left) and β (slashed line, scale on the right), where the interface width values were averaged over 30 different realizations of the random number generator. We can distinguish at least three regions of scaling. For an interval within $10 \lesssim t \lesssim 40$ the exponent β is close to the value 0.38 of the MBE universality class that appears when only the surface diffusion is considered (see discussion in [Barabási & Stanley, 1995], chapter 13). Afterwards, the coefficient β grows crossing the value 0.5 characteristic of random erosion up to a value of 0.6 at $t \approx 800$. Then the value of β seems to stabilize, or at least grows at a much slower pace.

Figure 7.6 shows three stages of the evolution of the surface at the times $t = 51$, $t = 306$, and $t = 1008$ indicated in Figure 7.5. The right column shows the power spectral density (PSD) computed by taking in to account all the $L = 2048$ cells of the surface. For $t = 51$ a noisy surface develops, and its PSD shows that there are non-negligible contributions of wave numbers over all the scales. For $t = 306$ the surface starts to develop ripple-

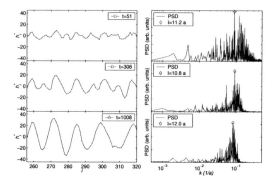

Figure 7.6: Three different stages of the evolution of the surface for the CMTHS model with $Y(\varphi) = 1$. Model parameters: $L = 2048$, $f = 0.5$ and $J/(k_B T) = 5$. The left column shows a 64 point-long portion of the height h_i^*, which is the height minus its average value at each time $h_i^* = h_i - \langle h_i \rangle$. The right column shows the power spectral density (PSD) computed with all the 2048 points of the surface versus the wave number k for each of the three stages. The PSD values are displayed in a linear scale.

like structures and the spectra of wave numbers gather around a maximum; when $t = 1008$ ripples with mean length $l \approx 12a$ are formed. The ripples are more regular and periodic than in the full CMTHS model. Due to the absence of the roughening generated by the non-linearities introduced by the yield function, the ripple regime with growing amplitudes persists for long time. Additionally, the ripple length is also increasing with time. This coarsening proceeds in such a way that smaller ripples are merged with their bigger neighbors, as it can be appreciated comparing the surface topography for times $t = 305$ and $t = 1008$ in Figure 7.6. We discuss further this effect in Section 7.5.

7.4   The extended model

The main motivation to modify the CMTHS model is to improve the features of the ripple regime. Because the instability is generated by the curvature-dependent erosion rule, it is natural to try to improve the curvature estimation of the original CMTHS model. Another issue that is addressed is to reduce the influence of the random fluctuations in the evolution of the surface. The surface diffusion rule is not modified. The model resulting from the combination of extension of the erosion rule and the original surface diffusion rule is called the *extended* model.

7.4.1 The erosion rule - extended model

We propose to replace the "box rule" illustrated in Figure 7.2 by a direct estimation of the derivatives, angles, and curvatures based on finite central differences around a selected site. The first derivative or gradient of the surface at a point i is $\nabla_i \equiv (h_{i-1} - h_{i+1})/2a$, where a is the width of the lattice width. The angle φ that the surface forms at this site is estimated by $\varphi = \arctan(\nabla_i)$. This angle corresponds to the incidence angle used in the formula (7.7). The second derivative is $\nabla_i^2 = (h_{i-1} - 2h_i + h_{i+1})/a^2$. The curvature at the site i can be estimated using the finite differences version of the curvature standard formula:

$$\kappa_i = \nabla_i^2(1 + (\nabla_i)^2)^{-3/2}. \tag{7.10}$$

Obviously, the non-linear character of this curvature estimation is source of deviations from the original CMTHS model.

Due to the discreteness of the height h_i, the values obtained with these formulas vary drastically from one site to the other. To attenuate this problem, the values for the angles and curvatures are computed not only for the site i, but also for its neighbors $i - 1$ and $i + 1$. The mean value of the angle for the site i is $\overline{\varphi_i} = (\varphi_{i-1} + \varphi_i + \varphi_{i+1})/3$, and the mean curvature is: $\overline{\kappa_i} = (\kappa_{i-1} + \kappa_i + \kappa_{i+1})/3$. This procedure takes into account the values of five sites $(h_{i-2}, h_{i-1}, h_i, h_{i+1}, h_{i+2})$, and provides a smoothed estimation of the angles and curvatures for the site i.

Taking into account these modifications in the algorithm, it is necessary to introduce two new parameters for estimating the curvature-dependent erosion probability p_κ: (i) the maximum of the positive curvature κ_{max} (the minimum κ_{min} is the negative of this value). In the case that the computed curvature κ_i were larger than κ_{max}, the algorithm assigns $\kappa_i = \kappa_{max}$. Similarly, when $\kappa_i < \kappa_{min}$ then $\kappa_i = \kappa_{min}$. (ii) The minimum of the curvature-dependent erosion probability is $p_{\kappa,min}$. The obtained values of the curvature κ_i are mapped by a linear transformation in such a way that for $\kappa_i = \kappa_{max}$, the curvature dependent erosion probability is $p_\kappa = p_{\kappa,max} \equiv 1$, and when $\kappa_i = \kappa_{min}$ then $p_\kappa = p_{\kappa,min}$.

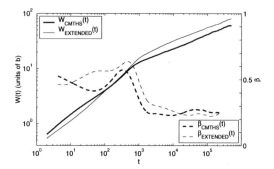

Figure 7.7: Comparison of the scaling of the surface width for the full CMTHS and full extended models (using $Y(\varphi) = 0.5 - 0.479\varphi^2 + 0.979\varphi^4$). For the CMTHS model: the surface width $W_{\mathrm{CMTHS}}(t)$ (bold continuous line, scale on the left) and the corresponding growth exponent β_{CMTHS} (bold dashed line, scale on the right). Parameters: $L = 2048$, $f = 0.5$, $J/(k_B T) = 5$. For the extended model: the width $W_{\mathrm{EXTENDED}}(t)$ (thin continuous line, scale on the left) and corresponding growth exponent $\beta_{\mathrm{EXTENDED}}$ (thin dashed line, scale on the right). The extra parameters of the extended model are the closest possible to the CMTHS model: $a = 1$, $b = 1$, $p_{\kappa,\mathrm{min}} = 1/7$, $\kappa_{\mathrm{max}} = 2$.

7.4.2 Scaling of the full extended model

In order to verify the connection of the extended model with the original CMTHS model, we selected a parameter set for the extended model that reproduces approximately the curvature values covered by the "box rule" of the CMTHS model. The evolution of the interface width for both models is compared in Figure 7.7. The system size is $L = 2048$, and the width values were averaged over 100 different realizations. The probability of invoking the erosion rule is $f = 0.5$ (the surface diffusion rule is invoked with probability $1 - f = 0.5$), and the constant associated with the diffusion is $J/(k_B T) = 5$. The extra parameters of the extended model are: $a = 1$, $b = 1$, $p_{\kappa,\mathrm{min}} = 1/7$, $\kappa_{\mathrm{max}} = 2$. For both cases, the same yield function $Y(\varphi) = 0.5 - 0.479\varphi^2 + 0.979\varphi^4$ is used. Thus, we are comparing the scaling behavior of the full versions of both models.

The scaling properties are similar, showing first a rough interface at early times, then an increase of the growth exponent β due to the onset of the instability that creates the ripples; and finally, a drop up to a value $\beta \approx 0.25$, which indicates the non-linear regime characterized by a rough surface. It is important to note that the scalings are not identical.

This is due to the improved curvature estimation and additional averaging procedures of the extended model. A precise identification of the limits of the ripple regimes for both models is difficult, because they depend on each realization and the criterion to distinguish between fluctuations and proper ripples. The ripple regime for the CMTHS model can be estimated to be $t \approx \{300, 1000\}$, while for the extended case the onset of the instability occurs earlier and the ripple regime lasts longer.

7.4.3   Scaling of the extended model with $Y(\varphi) = 1$

We claim that the extended model maintains the scaling properties of the CMTHS model (at least for the parameter set used in Figure 7.7). Therefore the extended model may be considered to describe the evolution of ion-sputtered surfaces. In Chapter 8, we apply the extended model with $Y(\varphi) = 1$ and a different parameter set to the simulation of pattern formation in the etching process. There we use the flexibility of the parameters of the extended model to generate ripples larger in size than the laser beam diameter. The obtained ripples are clearly distinguishable from the inherent fluctuations and roughness due to the stochastic character of the model.

In what follows we study the scaling properties and morphology evolution for the parameter set used in the LJE simulation presented in the 8. The probability of invoking the erosion rule is $f = 0.045$ (the surface diffusion rule is invoked with probability $1 - f = 0.955$) and the constant associated with the diffusion is $J/(k_B T) = 1$. For weighting the curvature-dependent erosion probability p_κ, the values $a = 20$, $b = 1$, $p_{\kappa,\min} = 0.1$, $\kappa_{\max} = 0.0004$ have been selected. The system size is $L = 2048$, and the width values were averaged over 100 different realizations.

Figure 7.8 shows the temporal evolution of the width $W_{\text{EXTENDED}}(t)$ and its growth exponent β_{EXTENDED} for this parameter set. The ripple regime persists much longer than in the cases presented in Figure 7.7, and the obtained ripples have a larger amplitude. For times $t \lesssim 150$, the exponent β is lower than the value 0.5, which is characteristic for random erosion. The strong increase of the value of β after $t \gtrsim 100$ is associated with the onset of the ripple regime. During the interval $10^3 \lesssim t \lesssim 10^4$, there is an oscillation of β, but the values remain relatively high. The origin and meaning of this oscillation is currently unknown, but similar oscillations of the interface width have been reported for a 2+1D discrete.model for sputtering erosion (see Figure 4 in [Hartmann *et al.*, 2002]). After $t \approx 10^4$ the growth exponent β shows a slow increase.

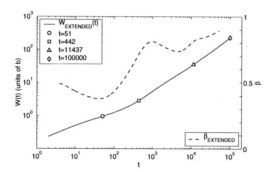

Figure 7.8: Temporal evolution of the interface width $W_{\text{EXTENDED}}(t)$ (continuous line) and growth exponent β_{EXTENDED} (dashed line) of the extended model with $Y(\varphi) = 1$. The morphology of the surface at times corresponding to $t = 51$ (○), $t = 442$ (□), $t = 11437$ (△), and $t = 10000$ (◇) is analyzed in Figure 7.9. Extended model parameters: $L = 2048$, $f = 0.045$, $J/(k_B T) = 1$, $a = 20$, $b = 1$, $p_{\kappa,\min} = 0.1$, $\kappa_{\max} = 0.0004$.

Figure 7.9 shows a portion of the surface for the four times that are indicated with symbols in Figure 7.8. The left column shows the height h_i^*, which is the height minus its average value at each time $h_i^* = h_i - \langle h_i \rangle$. The right column shows the corresponding *ensemble average* of the power spectral density (EA-PSD) computed with 30 realizations, and $L = 2048$. The PSD is represented in a logarithmic scale with arbitrary units, and the horizontal axis corresponds to the wave number k. A moving window average over 10 points is applied to the spectra, and the resulting *smoothed* ensemble average of the power spectral density SEA-PSD appears above the EA-PSD curve. The local maxima of the SEA-PSD (indicated by the (∇) symbols) can be used to estimate the mean value of the ripple length l of each stage.

In the rough surface corresponding to $t = 51$ (○), the integer values of the heights are noticeable. For $t = 442$ (□), the ripples start to appear, but their shape and length are highly irregular. Softer and larger ripples are found for times around to $t = 11437$ (△), and the length estimated by the SEA-PSD method is approximately $l = 23a$. It is observed that the ripple length increases with time due to the merging of ripples: small ripples are assimilated by contiguous larger ripples, which in turn develop a sinusoidal shape. For time $t = 10^5$ (◇), the ripple length is approximately $40a$. The increasing ripple length is also observed for the CMTHS with $Y(\varphi) = 1$ model (Section 7.3.4), as it would be discussed in Section 7.5. Analogous coarsening phenomena have been reported for

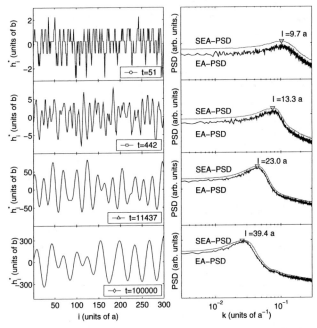

Figure 7.9: Four different stages of the evolution of the surface for the extended model with $Y(\varphi) = 1$, corresponding to the times referred in figure 7.8. The left column shows the height h_i^* (note that all vertical scales are different). The real system size is $L = 2048$, but only 300 points are shown. The right column shows the corresponding *ensemble average* of the power spectral density EA-PSD (lower curve) and the *smoothed ensemble average* of the power spectral density SEA-PSD (upper curve, shifted in the vertical direction for clarity of presentation). The local maxima of the SEA-PSD (indicated by the (∇) symbols) is used to estimate the mean value of the ripple length l of each stage.

experiments on ion sputtering [Habenicht *et al.*, 2002; Rusponi *et al.*, 1998], laser ablation [Georgescu & Bestehorn, 2004], and are also discussed from a theoretical point of view in [Muñoz-García *et al.*, 2005].

To summarize, the extended model with this particular parameter set generates ripples that are much larger and smoother than those of the CMTHS model. The curvature estimation algorithm of the extended model is sensitive to the small curvatures (positive or negative), which corresponds to longer ripple lengths. This has been accomplished using $f = 0.045$, which means that the surface diffusion rule is being applied 95.5% of the time, permanently minimizing the local fluctuations of the surface. A locally smooth surface is a necessary condition for an effective action of the curvature-dependent erosion rule. Thereby, fast growth of instabilities and in consequence larger ripples are obtained.

7.5   The continuum limit

Continuum representations are frequently more amenable to analytical treatment than discrete models. Describing the dynamics of a discrete model in the continuum limit as a partial differential equation (PDE) provides a essential representation of the phenomena, from which eventually, it is possible to get more information about the instabilities and non-linear properties of the system. The identification of this continuum equation could be used as a basis for the formulation of a more general PDE, from which quantitative results by numerical integration can be obtained. This should be performed taking in to account realistic values of the physical and chemical parameters of the materials and processes involved.

A master equation approach proposed by Vvendenky et al. [Vvedensky *et al.*, 1993] has been applied to the derivation of continuous equations for some discrete models [Chua *et al.*, 2005; Předota & Kotrla, 1996; Vvedensky, 2003a,b]. The regularization procedure expands the non-analytic quantities and replaces them with analytic ones. However, as pointed out by Předota and Kotrla, among other difficulties, the choice of a regularization scheme is ambiguous. Thus, the coefficients in the derived continuum stochastic equation cannot be determined uniquely.

Using the Vvedensky method, Lauritsen found a partial differential equation for the CMTHS model whose leading terms constitute a noisy Kuramoto-Sivashisnky (KS) equation [Lauritsen *et al.*, 1996]. This is in agreement with the fact that the numerical integration of a generic noisy KS equation has the same scaling behavior like the discrete CMTHS model (compare Figures 2(b) and 5 in [Cuerno et al., 1995]). We have already demonstrated that the scaling of the CMTHS and extended models are similar (at least for a specific parameter set), then we expect that the derivation of a continuum equation for the extended model is also related to the noisy KS equation. In what follows, we summarize the results presented in [Mora et al., 2005b].

7.5.1 From the transition rules to discrete stochastic equations

The discrete stochastic equation is derived for the extended model beginning with the master-equation description of the microscopic dynamics. We consider the birth and death type master equation [Van Kampen, 1981]

$$\frac{\partial P(H,t)}{\partial t} = \sum_{H'} W(H',H)P(H',t) - \sum_{H'} W(H,H')P(H,t), \tag{7.11}$$

where the array $H = \{h_1, h_2, \ldots\}$ corresponds to every surface configuration, and $P(H,t)$ is the joint probability distribution of having the configuration H at time t. $W(H,H')$ is the transitional probability per unit time from a configuration H to a subsequent configuration H'. Next, we define the first and second moments of the transition probability $W(H,H')$ as:

$$K_i^{(1)} = \sum_{H'} (h_i' - h_i)W(H,H') \tag{7.12}$$

$$K_{ij}^{(2)} = \sum_{H'} (h_i' - h_i)(h_j' - h_j)W(H,H'). \tag{7.13}$$

The master equation (7.11) is approximated by the Fokker-Planck equation through the usual Kramer-Moyal expansion (see [Vvedensky et al., 1993])

$$\frac{\partial P(H,t)}{\partial t} = -\frac{\partial}{\partial h_i}[K_i^{(1)}P(H,t)] + \frac{1}{2}\frac{\partial^2}{\partial h_i \partial h_j}[K_{i,j}^{(2)}P(H,t)], \tag{7.14}$$

where the sum over repeated indices is assumed. Then provided the system size is large and intrinsic fluctuations are not too large, the equivalent Langevin equation is written as

$$\frac{\partial h_i(t)}{dt} = K_i^{(1)} + \eta_i(t). \tag{7.15}$$

Here η_i represents white Gaussian noise with zero mean $\langle \eta_i(t) \rangle = 0$, and variance is given by the second transition moment

$$\langle \eta_i(t)\eta_j(t') \rangle = K_{ij}^{(2)}\delta(t - t'). \tag{7.16}$$

Equations (7.15) and (7.16) describe the temporal evolution of heights h_i at site i. For the extended model the transition probability is

$$W(H, H') = W_e(H, H') + W_d(H, H'), \tag{7.17}$$

where $W_e(H, H')$ and $W_d(H, H')$ are the transition probabilities per unit time for the erosion and surface diffusion rules respectively. The erosion transition probability reads

$$W_e(H, H') = \frac{f}{\tau}\sum_{i=1}^{L} p_\kappa Y_i \delta(h_i', h_i - b)\prod_{j\neq i}\delta(h_j', h_j), \tag{7.18}$$

where τ is the time scale. The parameters f, p_κ and Y_i have been defined in Sections 7.3 and 7.4. The erosion rule of the extended model is sensitive to the range of curvatures $[-\kappa_{max}, \kappa_{max}]$, and if the estimated curvature κ_i is outside this interval, κ_i is replaced for one of the two limits $-\kappa_{max}$ or κ_{max}. The curvature-dependent probability p_κ ranges from $p_{\kappa,min}$ to 1. The algorithm that assigns this probability can be represented by a smooth function

$$p_\kappa(\kappa_i) = \frac{1 + p_{\kappa,min}}{2} + \frac{1 - p_{\kappa,min}}{2}\tanh\left(\frac{\kappa_i}{\kappa_{max}}\right). \tag{7.19}$$

For $\kappa_i/\kappa_{max} < \pi/2$ the previous formula can be approximated for small surface gradients by

$$p_\kappa(\kappa_i) = \frac{1 + p_{\kappa,min}}{2} + \frac{1 - p_{\kappa,min}}{2}\nabla^2 h_i(1 - 3/2(\nabla h_i)^2) + \cdots. \tag{7.20}$$

The sputtering yield of Equation (7.7) can be approximated by

$$Y_i = y_0 + \frac{y_1}{a^2}(\nabla h_i)^2 + \frac{y_2 - 2y_1/3}{a^4}(\nabla h_i)^4 + \cdots \tag{7.21}$$

where $\nabla h_i = (h_{i+1} - h_{i-1})/2$. The transition probability for the surface diffusion rule reads:

$$
\begin{aligned}
W_{\mathrm{d}}(H, H') &= \frac{1-f}{2\tau} \sum_{i=1}^{L} [w_i^+ \delta(h_i', h_i - b)\delta(h_{i+1}', h_{i+1} + b) \\
&\quad + w_{i+1}^- \delta(h_i', h_i + b)\delta(h_{i+1}', h_{i+1} - b)] \times \prod_{j \neq i, i+1} \delta(h_j', h_j).
\end{aligned} \quad (7.22)
$$

Here, $w_i^{\pm}$ are the hopping probabilities defined in Equation (7.9). For example, w_i^+ can be approximated by:

$$
w_i^+ = \frac{1}{1+q}\left(1 - \frac{q\gamma}{1+q}\nabla^3 h_i + \cdots\right), \quad (7.23)
$$

where $q = \exp(6J\beta)$, $\gamma = 2J\beta a^2/b$, $\beta = 1/(k_B T)$. In the small gradient approximation, and neglecting higher order terms, the transition moments for the extended model can be simplified (see detailed description in [Mora *et al.*, 2005b])

$$
K_i^{(1)} = -\frac{b}{\tau}\left(f p_\kappa Y_i + \frac{(1-f)q\gamma}{(1+q)^2}a^2 \nabla^4 h_i\right) \quad (7.24)
$$

$$
K_{ij}^{(2)} = \frac{b^2}{\tau}\left(f p_\kappa Y_i \delta_{ij} - \frac{1-f}{1+q}\nabla^2 \delta_{ij}\right) \quad (7.25)
$$

where $\nabla^2 \delta_{ij} = \delta_{i-1,j} - 2\delta_{ij} + \delta_{i+1,j}$. The expressions for p_κ and Y_i are given by (7.20) and (7.21).

7.5.2 A Langevin equation for the extended model

A coarse graining procedure of the discrete stochastic equations (7.15) and (7.16) is a non-trivial task and there are few situations for which it has been done with mathematical rigor. Thus, in practice, coarse graining of the discrete Langevin equations is replaced by smoothing out the height difference in the limit of lattice constant $a \to 0$, as follows

$$
h_{i\pm 1}(t) - h_i(t) = \sum_{n=1}^{\infty} \frac{(\pm a)^n}{n!}\frac{\partial^n h}{\partial x^n}\bigg|_{x=ia}, \quad (7.26)
$$

such that $h_i(t) = h(x = ia, t)$. Because we are concerned with the form of the continuous Langevin equation for $h(x, t)$, we insert the expressions (7.24) and (7.25) into the discrete Langevin equations (7.15) and (7.16). Therefore we can identify the leading terms of the expansion of (7.26):

$$\frac{\partial h}{\partial t} = v_0 + \nu \nabla^2 h - K \nabla^4 h + \lambda (\nabla h)^2 + c_1 \nabla^2 h (\nabla h)^2 + \cdots + \eta(x, t), \qquad (7.27)$$

where $h = h(x, t)$, and v_0 is the vertical mean velocity of the moving interface. The terms related with the variables ν, K, and λ together with the noise term $\eta(x, t)$ constitute the noisy Kuramoto-Sivashinsky (KS) equation. In consequence, the linear stability analysis of the KS equation performed in the previous chapter should also hold here. The expression (7.27) only shows the lowest order terms of an infinite series, which agrees with the fact that for any discrete model there is, in principle, an infinite number of nonlinearities [Muraca *et al.*, 2004]. Because the coefficients of the expansion ν, K, λ, c_1, ... can be expressed in terms of the set parameters of the discrete model, comparisons could be drawn between the discrete model and the numerical integration of Equation (7.27).

The CMTHS and extended models (even in their $Y(\varphi) = 1$ versions) introduce non-linear behavior which differs from the dynamics of the non-linear term $\lambda (\nabla h)^2$ of the Kuramoto-Sivashinsky equation. Namely, an exponential growth of the instability is not possible within both discrete models (see discussion in Reference [18] of Cuerno *et al.* [1995]). Furthermore, the curvature estimation in both discrete models is based in a truncation of the values of the estimated local curvature, which introduces non-linear behavior. In particular, it is important to emphasize that in the extended model, the curvature estimation given by (7.10) *is* non-linear and should correspond to high order terms of the gradient expansion (7.27).

Apart from the possibility of performing linear stability analysis, the importance of writing explicitly a continuum equation that corresponds to the extended model is to identify the cause of the deviation from a behavior only ruled by a Kuramoto-Sivashinsky equation. For example, we have found that the extended model belongs to a category of pattern forming systems characterized by a permanent coarsening of the pattern wavelength [Politi & Misbah, 2004]. We claim that this phenomena can be caused by the non-linear term $c_1 \nabla^2 h (\nabla h)^2$ and other high-order terms found in Equation 7.27 (see [Raible *et al.*, 2000]).

7.6   Discussion

The extended model improves the pattern formation features of the original CMTHS model. This has been accomplished by means of a more accurate estimation of the curvature on each lattice site. Because the curvature values are averaged over nearest neighbors, the effect of random fluctuations of the height is reduced, and the curvature-dependent erosion rule generates instabilities more easily. In consequence, the extended model produces a ripple regime that lasts longer and whose ripples develop longer wavelengths. These features are necessary for the application of extended model presented in the next chapter.

Following the methods proposed by Vvedensky and others, we found that under the small gradient approximation, the extended model can be related with a continuum Langevin equation, which is a modified noisy Kuramoto-Sivashinsky equation (KS) (Equation 7.27). The existence of a ripple regime followed by a roughening of the surfaces is obviously explained by the presence of the KS terms. Whereas, the coarsening of the ripples (which is a deviation from the KS behavior) can be attributed to non-linearities introduced by high order terms.

Assuming that the pattern formation phenomena appearing in ion sputtering presents universality characteristics, in next chapter, we formulate an application of the extended model for ripple formation in the LJE experiment.

Chapter 8

The LJE simulation

In this chapter we combine together results from previous chapters, to simulate ripple regimes appearing in kerfs structured in the experiment. The *LJE simulation* consists in the application of the erosion and surface diffusion rules of the extended model within a moving probability distribution, which is proportional to the temperature field produced by laser absorption. This *etching probability distribution* is derived from the thermally thin layer approximation presented in Chapter 5, and simulates the fact that etching occurs only above a certain temperature threshold in a region which is broader than the laser spot.

The performed simulations provide a good qualitative comparison with the experiments of kerfs structured with varying feed velocity and varying laser power, where ripple regimes appear. Furthermore, the simulations provide a test of the working hypothesis about the relationship between laser power and ripple length proposed in Chapter 6. In addition, the process of creation of a single ripple is analyzed providing hints about the conditions for the onset of such instabilities.

8.1 Discrete models for etching

Extensive literature exists on the simulation of thermally activated chemical etching by means of discrete models. For example, a lattice model initially proposed for dissolution kinetics of minerals, and further developed by Poupart *et al.*, [Poupart & Zumofen, 1992], is based on temperature dependent microscopic dissolution rules. The dissolution probabilities were assumed to depend on local structure and temperature. The probability of a cell being removed is proportional to $exp(-nw)$, where n is the number of the cell's nearest neighbors and $w = 1/T$ where T is the temperature. The resulting surfaces present overhangs. As the parameter w varies, the model shows changes in its universality class.

Sapoval *et al.,* [Sapoval *et al.,* 1998] proposed a lattice model where a random probability between 0 and 1 is assigned to each cell representing the resistance to the etchant. The algorithm corrodes stepwise the surface in contact with the etchant solution, revealing a fractal or self-affine morphology when the chemical reaction stops. There is qualitative agreement with the experiment, and the model is suitable to be used for describing the dynamics at an atomic scale. This model also presents overhangs.

Concerning solid-on-solid models, Fernandes *et al.,* proposed a $2 + 1$ dimensional model for anodic dissolution Fernandes *et al.* [1993]. Within this model, the dissolution probabilities were proportional to $\min(1, exp(V - nE))$, where n is the number of occupied neighbors and E is related with the activation energy needed to break a bond between a filled site and one of its filled neighbors. The parameter V accounts for the biasing of the reaction rates due to the potential. In this model the surface roughness increases with the applied potential. More recently, Mello *et al.,* [Mello *et al.,* 2001] presented a simple atomistic model for dissolution, which mimics the etching of a crystalline solid by a liquid. The numerical simulations indicate that the model satisfies the Family-Viscek scaling and belongs to the KPZ universality class [Barabási & Stanley, 1995; Kardar *et al.,* 1986]. It can be applied to simulate the roughening dynamics of the LJE, but it does not contains pattern formation properties.

8.2   The extended model applied on etching

While the CMTHS model was derived for ion sputtering, it can be stated that the erosion and surface diffusion rules are of generic nature. Our extended model enhances the pattern formation features of the CMTHS model, allowing for its application to other physical systems. We claim that the curvature dependent erosion and the thermally activated surface diffusion mechanisms can be used also for explaining ripple formation in the structuring of ripples by LJE. The justification is based on experimental evidence, which shows that under certain conditions, "valleys" (portions of surface with positive curvature) are preferentially etched as compared to "peaks (portions of surface with negative curvature). This can be explained as a consequence of differences in the absorption of

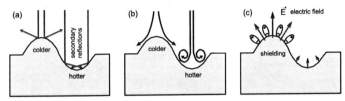

Figure 8.1: Proposed mechanisms to justify that, for the LJE experiment valleys are eroded more efficiently than peaks. (a) Secondary reflections. When the laser beam illuminates a peak, the reflected rays diverge, while reflected rays from a valley can converge to a point above the valley, and then secondary reflections can occur. Hence, the valley will be hotter than the peak, and in consequence, more erosion will occur there. (b) Hydrodynamic conditions. The flow of etchant and its convective cooling effect is stronger on the peak where the flow is more uniform compared to the valley, where stagnation of flow and eddies are more probable. Therefore, a "hot" valley is eroded faster than a "cold" peak. (c) In the case of electropolishing at nanometer scale [Yuzhakov *et al.*, 1997, 1999], preferential absorption of organic molecules on the peak, due to its enhanced electric field produces a shield that prevents large etching rates. Therefore, valleys are etched more efficiently than peaks (see details in the text).

laser energy and the resulting heat processes. When the laser beam is acting on a valley, the reflected rays converge to a region above the valley, and then they can eventually produce secondary reflections inside (see Figure 8.1(a)). This means that the surface of the valley can effectively absorb more laser radiation, more heat is produced, and in turn the etching rate is enhanced. In the case of peaks, reflected rays are dispersed in all directions and there are no secondary reflections. Regarding the cooling effect of the etchant jet due to heat convection, peaks are cooled more efficiently than troughs where stagnation of the flow, and even eddies, are more probable to appear (see Figure 8.1(b)). Therefore, the preferential heating and etching within troughs result in further increase of the local curvature, which in turn enhances the secondary reflections and heating due to poorer convective transport of heat by the etchant flow. This mechanism can be considered the origin of the unstable erosion mechanism.

Although the scales and the involved processes are different to those of LJE, it is worthy to mention the case of nanoscale pattern formation during electrochemical dissolution of anodically polarized aluminum in a variety of electrolyte solutions (see [Yuzhakov *et al.*, 1997, 1999]). A non-uniform interface creates a higher electric field at the peaks

(a)   (b)

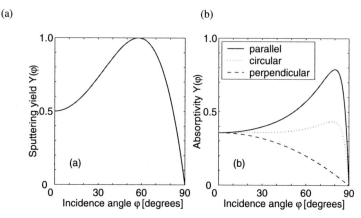

Figure 8.2: (a) The sputtering yield dependence with the angle of incidence. The function used in the simulations is $Y(\varphi) = 0.5 + 0.979\varphi^2 - 0.479\varphi^4$ and $Y(0°) = 0, Y(57.3°) = 1, Y(90°) = 0$. (b) Absorptivity of polarized light versus incidence angle for a flat iron surface (based on [Yao, 2000]). Plane of polarization parallel to the incidence plane (continuous line), circular polarization (dotted line), plane of polarization parallel to the incidence plane (dashed line). The coefficient of refraction is $n = 3.81$ and the attenuation coefficient is $k = 4.44$.

relative to the valleys. Therefore, organic molecules, with slightly higher polarizability than water, will absorb preferentially onto the peaks due to the larger electrical field. The absorbed molecules physically and chemically block the interface at peaks, thereby reducing the dissolution reaction. The competition of these two mechanisms is balanced and a stable pattern form [Guo & Johnson, 2003]. This phenomenon deviates from the normal electropolishing techniques, where etching rates at peaks are higher than etching rates on valleys and the surface becomes smooth.

After having presented a justification for the use of the rules of the extended model, it is necessary to find an analogy of the sputtering yield function $Y(\varphi_i)$. This can be accomplished by taking into account that the absorption of polarized light by metallic surfaces has similar functional dependence as the sputtering yield function (specifically in the case of the electric field parallel to the plane of incidence or even circular polarization, see Figure 8.2) .

8.3 The etching probability distribution

The etching reaction is driven by the field of temperature produced by absorption of laser light and the permanent transport of reactants and products performed by the etchant jet. Another effect of the etchant jet is to reduce the temperature of the surface by heat convection. In the analyzed experiments, the etchant jet velocity is constant; thus for our purposes, the resulting cooling effect can be considered to reduce the temperature field. In our simplified model, we assume that the etching rate at each point is proportional to the temperature and is not limited by transport processes.

Given the thickness of the metallic sample ($\approx 200 \mu m$) and the relatively large dwell times of the laser beam (of the order of seconds), the back surface of the sample reaches approximately the same temperature as the front surface on which the radiation is incident. Therefore is reasonable to use the thermally thin layer approximation of the field of temperature presented in Section 5.3 to define the etching rate in each point. Regarding the fact the diffusion of heat occurs much faster than the characteristic dwell times used in the experiment, it can be assumed that the estimated field of temperature is stationary in the reference frame of the moving laser.

The moving Gaussian distributed intensity of the laser can be expressed by means of:

$$F(r, t) = \frac{F_0}{\sigma\sqrt{2\pi}} e^{-(r - v_f t)^2 / (2\sigma^2)} \tag{8.1}$$

where r is the spatial coordinate, t the time, σ corresponds to the standard deviation and v_f is the feed velocity. The factor F_0 is a constant amplitude. Regardless of variations related to the choice of the various thermal constants and parameters, a generic profile of the temperature field can be identified (see details in Chapter 5). The curve shown in Figure 8.3 is proportional to such a profile, which can be described to be similar to a Gaussian shape within the laser spot area, while for larger distances it decays as $-log(r)$.

We define the *etching probability distribution* P_E of applying the erosion and diffusion rules on each site to be directly proportional to the estimated temperature profile (see Figure 8.3). The factor of proportionality is chosen in such a way that for maximum power, the probability at the central point is one, $P_E(r = 0) \equiv 1$. Introducing an additional factor between 0 and 1 into the P_E distribution, it is possible to simulate laser beam powers between 0% and 100% of the maximum power.

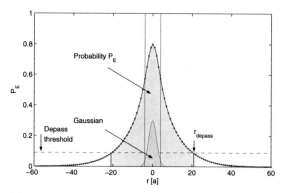

Figure 8.3: Probability P_E is proportional to an estimation of the temperature field produced by a Gaussian beam with standard deviation $\sigma = 2a$ for 80% power and a depassivation threshold of 0.09. The etching probability P_E determines how frequently the erosion and diffusion rules are applied, and is used to simulate the joint action of the moving laser beam and etching jet for different powers and feed velocities. In order to compare the sizes of laser beam and the etching front, a Gaussian distribution with the standard deviation ($sigma = 2a$) of the laser beam is depicted. The vertical lines are located at $r = \pm 2\sigma$ defining the diameter of the laser spot d_{sim} used in the simulations.

In order to simulate the fact that etching occurs only above a certain temperature, a depassivation threshold is introduced and P_E is defined to be zero below it (see the slashed horizontal line in the figure). Therefore, the etching probability distribution has a finite diameter d, which depends on the power and the depassivation threshold. In the experiment, the temperature above which depassivation occurs is not known precisely and depends strongly on the etchant concentration. For the simulations in this chapter, a depassivation threshold of 0.09 has been selected with the criterion of obtaining a diversity of etching probability distributions when the power is varied.

Figure 8.3 shows the P_E distribution for 80% power. The resulting diameter of the P_E distribution is $d = 42a$. In order to compare this diameter with the effective laser spot diameter $d_{sim} \equiv 4\sigma = 8a$, the corresponding Gaussian distribution is also depicted in the figure. The resulting etching front is approximately five times larger than the laser beam diameter which is comparable with the situation in the experiment.

In the simulation, the metallic sample is kept fixed while the etching probability distribution moves with velocity v_f. This feed velocity is measured in units of a per step of the algorithm. The algorithm randomly selects a cell within the etching probability distribution by means of a rejection Monte Carlo technique used normally for integration (see [Press et al., 1992, Section 7.3]). Within the etching front, a combined action of erosion and surface diffusion rules occurs during a characteristic dwell time d/v_f. Depending upon v_f and the power, the etching front will eventually develop instabilities, giving rise to ripple structures. After the etching front leaves a region, the surface has acquired a topography that, depending on model parameters, corresponds roughly to one of the stages analyzed in Figure 7.8.

In the experiment, the absorption of polarized laser light depends on the incidence angle at each point. This can be accounted by the absorption function $Y(\varphi)$ (formerly named sputtering function, see figure 8.2(b)). Depending on the geometry of the surface illuminated by the laser, variations of the absorption of laser energy and generated heat must imply variations in the field of temperature. This effect should be noticeable during ripple formation due to the permanent variations of slope. However, we assume that this angle dependence of the absorption of light at the laser spot can be neglected in a first approximation. Therefore, we use the extended model with $Y(\varphi) = 1$.

In order to include this effect in further developments of the LJE simulation, it will be necessary to compute an average of the light absorption at the spot for each step, and then

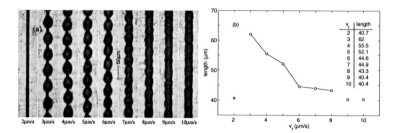

Figure 8.4: (a) Optical microscope images showing a series of kerfs structured with increasing feed velocities v_f from 2μm/s to 10μm/s analyzed previously in Chapter 3 (b) Dependency of the ripple length on the feed velocity (○ symbols). For feed velocities $v_f = 2, 9, 10\mu$m/s, almost uniform kerfs are produced. However, an oscillation of $40\mu m$ related with the xy-stage (□ symbols) can be identified within the uniform kerfs.

the corresponding etching probability distribution. Depending on the chosen absorption functions, this ever-changing etching probability distribution can introduce additional oscillations in the system.

8.4 Simulations with varying feed velocity

Proper selection of model parameters allows one to perform simulations that can be compared with experiments. In Figure 8.4(a), we recall the kerfs structured with varying feed velocities analyzed in Chapter 3. The main feature is a ripple regime, where the ripple length decreases monotonically with the feed velocity v_f as shown in Figure 8.4(b).

Figure 8.5 shows the simulated kerfs at feed velocities $v_f = 1, 2, 6$ and $32 \times 10^{-6}(a/\text{step})$ using the etching probability distribution for 80% power presented in Figure 8.3. The applied extended model parameters are: $Y(\varphi) \equiv 1$, $f = 0.045$, $a = 20$, $p_{\kappa,\min} = 0.1$. In the profile corresponding to $v_f = 1 \times 10^{-6}a/\text{step}$, at the beginning there is a transient oscillation of the height (left hand side), after which the etching front becomes stable. The resulting bottom of the kerf is rough, but can be considered flat on average. For feed velocities between $1 < v_f \lesssim 2 \times 10^{-6}a/\text{step}$, two or more transient oscillations can appear, but a uniform etching front is formed afterwards. This can be compared with uniform kerfs from the experiment, which present a transient region with a topography

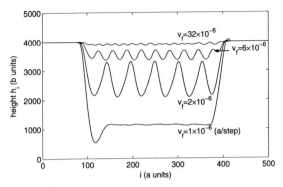

Figure 8.5: Profiles of the kerfs corresponding to feed velocities $v_f = 1, 2, 6$ and $32 \times 10^{-6} a/$step structured with the temperature-related probability P_E of Figure 8.3. The horizontal axis is the site number i measured in units of a (width of the cell), and its corresponding height is measured in units of b (height of the cell). Extended model parameters: $Y(\varphi) \equiv 1$, $f = 0.045$, $a = 20$, $p_{\kappa,\min} = 0.1$, $\kappa_{\max} = 0.0004$.

deeper and rougher than the rest of the kerf.

Returning to the simulations, the fact that uniform kerfs appear at small feed velocities can be explained in the context of the extended model as follows. Smaller feed velocities imply longer dwell times, and therefore the etching fronts become more tilted. Larger slopes mean that in the foremost part of the etching front, namely the corner formed by the etched and non-etched surfaces, the values of the negative curvature become progressively smaller (larger in absolute value). Considering that the erosion rule detects only a curvature range from $(\kappa_{\min}, \kappa_{\max})$, it is possible that the curvature at the mentioned region is always smaller than $\kappa_{\min}$, and in consequence the erosion rule acts permanently with probability $p_e = p_{\kappa,\min}$. This is equivalent to always have curvature-independent erosion combined with surface diffusion, and in turn, there are no conditions for instabilities in the foremost part of the etching front, and the kerfs become uniform. The transient oscillations are an exception to the rule, because at the very beginning of the simulation when the P_E distribution just creates hollow on the flat surface, the curvature values are indeed covered by the erosion rule, allowing for instabilities and ripples. When the etching front becomes deeper, the transient oscillations are quickly damped.

For feed velocities equal to and larger than $2 \times 10^{-6} a/$step, the obtained periodic ripples

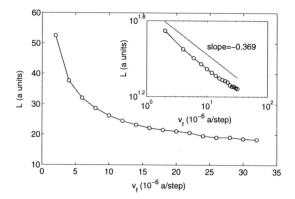

Figure 8.6: Ripple length as a function of the feed velocity v_f corresponding to the parameter set used in Figure 8.5. For feed velocities less than $2 \times 10^{-6}a/\text{step}$, there is no ripple formation (see the profile corresponding to $v_f = 1 \times 10^{-6}a/\text{step}$ in Figure 8.5). The insert shows the same curve plotted in a *log-log* scale.

present an amplitude of the same order of the mean depth. The ripple length decreases with increasing feed velocity. Larger feed velocities mean shorter dwell times and the kerfs become shallower. A short dwell time means that the resulting topography at the bottom of the kerf can be related with early rough stages of the evolution shown in Figures 7.8 and 7.9 of the previous chapter. For feed velocities greater than $32\mu\text{m/s}$, the ripples have short lengths and amplitudes and they are so irregular that they can be considered roughness of the surface.

In order to evaluate the dependence of the ripple length with the feed velocity, we examined the range from $1 \times 10^6 a/\text{step}$ to $32 \times 10^6 a/\text{step}$. An ensemble average of the Fourier power spectral density (PSD) over five realizations of the bottom of the kerf (each 16384 points long), allows us to estimate the ripple length for each feed velocity.

Except for velocities larger than $32a \times 10^{-6}(a/\text{step})$, the ripples are periodic and the ripple length is defined almost unambiguously. Figure 8.6 shows a monotonically decreasing curve which agrees qualitatively with the experimental curve shown in the 8.4(b). The insert of Figure 8.6 shows the same curve plotted in a log-log scale in order to check a power law behavior. The fit $L \propto v_f^{-0.369}$ is poor, and therefore the hypothesis of a power law is inconclusive.

(a)   (b)

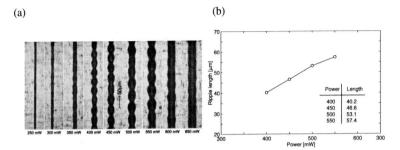

Figure 8.7: (a) Reproduced from Rabbow [Rabbow *et al.*, 2005]. Optical microscope images showing a series of kerfs structured with increasing powers from 250 mW to 650 mW (etchant 5M H_3PO_4, feed velocity 6μm/s, etchant jet velocity 190 cm/s). The kerfs for 250, 300 and 350 mW are the result of a uniform etching front which widens with power. For powers greater than 350 mW, instability in the etching front appear and the obtained ripples seem to be a product of thermal runaways. (b) Ripple length of the kerfs from 400 to 550 mW ($\circ$ symbols). The kerfs corresponding to powers $\lesssim 350$ mW and $\gtrsim 600$ mW are considered uniform.

8.5 Simulations with varying power

In order to compare with the simulations, in Figures 8.7(a) and (b) we recall the experiment of kerfs performed with varying powers, as described in Chapter 6. The main feature is the presence of periodic ripples in the range from 400 mW to 550 mW with increasing characteristic length (see Figure 8.7(b)).

The simulation of kerfs for different laser powers requires the computation of the corresponding etching probability distributions P_E. For powers 20%, 60% and 100% and fixed depassivation threshold of 0.09, Figure 8.8 shows their corresponding P_E distributions. Certainly, their diameters increase with power: $d = 16a, 36a$ and $46a$ respectively. In order to compare these diameters with the laser spot diameter used in the simulations $d_{sim} \equiv 4\sigma = 8a$, its corresponding Gaussian distribution is also shown in the figure.

Before we examine a whole series of simulated kerfs, it is worth to have a close look at the creation process of a single ripple as illustrated in Figure 8.9. The etching probability distribution P_E, corresponding to 60% power shown in Figure 8.8, is used to etch a kerf on an initially flat surface with feed velocity $2 \times 10^{-6}a$/step. On each frame of the figure,

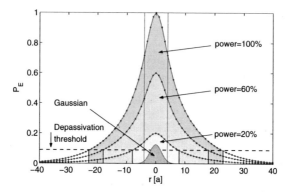

Figure 8.8: Etching probability distribution P_E associated with the estimation of the temperature field produced by a Gaussian beam with standard deviation $\sigma = 2a$ for beam powers 20%, 60% and 100%. The etching probability P_E is used to simulate the joint action of the moving laser beam and etching jet, and determines how frequently the erosion and diffusion rules are applied. A Gaussian distribution with the standard deviation ($sigma = 2a$) of the laser beam is depicted. The vertical lines are located at $r = \pm 2\sigma$.

a horizontal gray scale bar represents the moving etching probability. The vertical dotted lines define the region where the etching front is active. The extended model parameters are the same as used in the previous section. Typically, a valley is created at the forefront of the moving P_E from a small, but growing, depression on the surface. When the center of P_E passes through a valley, the rate of erosion increases, and the valley grows as long as the rearmost part of the P_E acts. The peaks between the valleys are eroded at a lower rate due to their negative curvature. In summary, the local and temporal action of the etching probability distribution works as an amplifier of small instabilities on the surface.

Figure 8.10 shows the profiles of the kerfs structured with etching probability distributions P_E corresponding to powers ranging from 20% to 100% and constant feed velocity $2 \times 10^{-6}a$/step. The parameters of the extended model are the same as used previously. When power is increased, the diameters and amplitude of the corresponding P_E distributions grow proportionally. Larger diameters d mean longer dwell times d/v_f and together with larger etching probability, the resulting topographies correspond to later stages of the evolution of the extended model.

As illustrated in Figure 8.8, the etching probability distribution corresponding to 20%

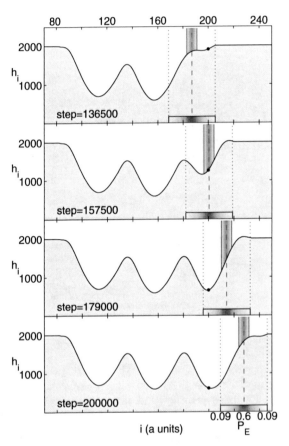

Figure 8.9: Four stages of the formation of a single ripple. The etching probability distribution P_E corresponding to 60% power shown in Figure 8.8 is represented by the gray-scale bar at the bottom of each frame. The standard deviation of the moving Gaussian beam is $\sigma = 2a$. Its center is represented by a vertical dashed line. The vertical continuous lines are located at $\pm 2\sigma$ in a co-moving system. A marker illustrates the evolution of a point on the surface at a fixed position $i = 200$. The feed velocity is $2 \times 10^{-6} a/\text{step}$. Extended model parameters are: $Y(\varphi) \equiv 1$, $f = 0.045$, $J/(k_B T) = 1$, $a = 20$, $p_{\kappa,\text{min}} = 0.1$, $\kappa_{\text{max}} = 0.0004$.

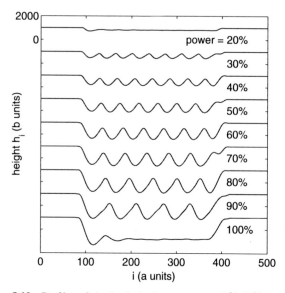

Figure 8.10: Profiles of the kerfs for beam powers $20\%, 30\%, \ldots, 100\%$ structured with the temperature-related, etching probability distributions P_E shown in Figure 8.8. All kerfs are structured with the same feed velocity $2 \times 10^{-6} a/\text{step}$. For 20% power, the kerf is shallow and rough, while for powers from 30% to 90%, a ripple regime with increasing ripple length appears. For 100% beam power, the bottom of the kerf again becomes approximately flat with small roughness. The extended model parameters are the same as in the previous figure.

power has a small diameter and overall amplitude. Therefore, when this distribution is scanned along the surface, the resulting topography corresponds to early stages of the evolution where the ripple regime is not yet reached. In consequence, the kerf becomes shallow and rough. For powers ranging from 25% to 90%, the etching probability distributions grow in diameter and amplitude, and stages with very regular ripples are obtained. Regarding the phenomenon of ripple coarsening found in the extended model, it is natural to obtain ripples with increasing length when the power is increased.

For beam power 100%, after a few initial transient oscillations, the bottom of the kerf becomes approximately flat with small roughness. As discussed in the previous section, this is due to the fact that for large powers and long dwell times, the negative curvatures within the etching forefront are always outside the range of curvatures that can be distinguished by the erosion rule. Therefore, the algorithm assigns the minimum of the erosion probability $p_{e,\min}$. In consequence, the curvature related instabilities cannot be generated nor amplified, and the kerfs become uniform. As explained in the previous section, transient oscillations appear only in the transition from a flat surface to a fully developed etching front.

An ensemble average of the Fourier power spectral density (PSD) over 10 realizations of the bottom of the kerf (each 16384 points long), allows us to estimate the ripple length for each power between 25% to 90%. It has been found that for each power, there is a definite ripple length with negligible uncertainty. Figure 8.11(a) shows how this ripple length increases approximately in a linear way with the power (○ symbols). The diameter d of the etching probability distributions (depicted with the △ symbols) also grows with the power. In order to have a reference, the diameter of the laser spot $d_{sim} = 8a$ used in the simulation is indicated in the figure by the horizontal line.

These results are in agreement with our working hypothesis for the experiment: for constant feed velocity, the average diameter of the etching front (uniform or unstable) increases with the power due to more heat and, when there are conditions for instabilities, the length of the resulting ripples is proportional to the average diameter of the etching front. In the context of the simulation, this can be expressed as follows: when instabilities and ripples appear, their characteristic length grows proportionally with the diameter of the etching probability function P_E.

In order to compare our model with the experiment, Figure 8.11(b) shows ripple lengths corresponding to powers 400, 450, 500 and 550 mW of Figure 8.7. The ripple length

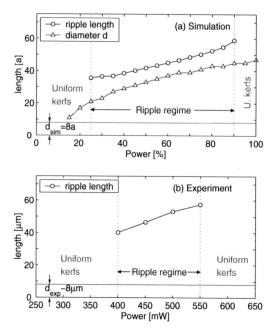

Figure 8.11: (a) Simulation. Increasing ripple length (○ symbols) for beam powers $25, 30, 35, \ldots, 90\%$ corresponding to the same parameter set as in the two previous figures. For powers 15% and 20% the kerfs are shallow and uniform. For 95% and 100%, the kerfs become uniform after a few transient oscillations. The diameters d (△ symbols) of the corresponding etching probability distributions P_E increase with power. (b) Experiment. Ripple length (○ symbols) of the kerfs shown in Figure 6.2 correspond to powers 400-550 mW. The kerfs corresponding to powers $\lesssim 350$ mW and $\gtrsim 600$ mW are considered uniform. For both (a) and (b), the region between the two dotted vertical lines approximately define the power interval where periodic ripples appear.

increases proportionally with the laser power. The diameter of the laser spot $d_{exp} \approx 8\mu m$ used in the experiment is indicated in the figure by the horizontal line. Of course, four points are not enough to draw conclusions about this dependency. In summary, the simulations show a qualitative agreement of our hypotheses with the experiment: the etching front covers a region much wider than the laser spot, and the resulting characteristic ripple length is proportional to the diameter of the active etching zone.

8.6 Discussion

The extended discrete model applied to the simulation of kerf formation qualitatively reproduces some of the main characteristics of microstructures produced by the LJE experiment. This is accomplished by applying the erosion and diffusion rules within a co-moving etching probability distribution P_E, which is proportional to an estimated temperature field produced by the laser spot. This distribution allows one to model the fact that etching occurs in a region much broader than the laser spot. Taking advantage of the probabilistic nature of the kinetic Monte Carlo method, it is possible to simulate the laser power distribution, depassivation threshold, and feed velocity.

This is a phenomenological model focused on the pattern formation phenomenon. The main advantage of the LJE simulation is the flexibility provided by its numerous parameters, which allowed us to perform simulations that can be compared qualitatively with the experimental results. In particular, variations of the feed velocity (maintaining the power constant) reveal a regime with an unstable etching front and in turn, regular ripples in the kerfs. As in the experiment, the ripple length decreases monotonically with increasing feed velocity. When the feed velocity is constant and the power is increased, a regime with ripples is found. For a certain interval of powers, ripples with increasing length appear to qualitatively agree with experimental results. More importantly, the working hypothesis about the proportionality between power, average size of the etching front, and resulting ripple length is supported by the simulations.

On the other hand, apart from the spatial scales that can be coarsely ascertained in the heat transport analysis, up to now it is not possible to compare the spatial lengths, velocities, powers or etching rates with those from experiments. A logical step would be to find a criterion for a correspondence between the lengths of the experiments and simulations, and then to infer the remaining equivalences from systematic experimental data. Identification of adimensional numbers with physical meaning will help to establish further links between experiments and simulations.

Finally, for a better comparison with the experimental results a further obvious step is the generalization of the model to 2+1 dimensions. Advances in this direction are shown in the appendix, where uniform kerfs and kerfs with ripples appear depending on the feed velocity.

Chapter 9

Conclusions

Ultraprecision microstructuring is one of the main goals of the current scientific and technological development. Control of the topography of the structures and quality of the surfaces is determinant for electronic, optical, and micro-electro-mechanical applications. Etching of silicon has been an extensively studied and established technique due to its obvious relevance in the semiconductor industry. In comparison, etching of metallic samples at micrometer scales is an open field with a lot of application potential, but still requires considerable research effort. Laser melting techniques are not suitable for microstructuring due to the effects of the high temperatures involved. The laser-induced jet-chemical etching (LJE) of metals has been proposed as a novel alternative for microstructuring of metallic surfaces [Metev *et al.*, 2003; Nowak & Metev, 1996; Rabbow *et al.*, 2005; Stephen *et al.*, 2002, 2004]. In the LJE experiment two type of structures are obtained: uniform and rippled kerfs. In both cases the walls and bottoms of the kerfs present rough surfaces. This is a natural consequence of the stochastic nature of laser light absorption, hydrodynamics of etchant, and chemical etching reactions. The kerfs with ripples are unwanted for many applications, and thus a better understanding of this phenomenon could provide hints of mechanisms to suppress it.

The main result of this work is that the *extended model* and its application in the *LJE simulation* provide a simple framework to simulate pattern formation appearing in the LJE experiment [Mora *et al.*, 2005a,b]. In spite of a lack of detailed microscopic knowledge of the various processes involved, universality assumptions have been successfully applied to qualitatively reproduce ripple regimes appearing in kerfs structured when the feed velocity or laser power are varied. This has been accomplished representing the various chemical and physical processes by only two competing physical mechanisms: curvature-dependent erosion and thermally-activated surface diffusion. These mechanisms are incorporated in the extended model as rules to be applied on cells of the 1+1 dimensional lattice that represents the surface. The temporal evolution of the model is driven by a kinetic Monte-Carlo method, which provides a means to represent in a probabilistic sense some of the main physical parameters of the experiment: laser power, feed velocity, and temperature field. In what follows, we discuss the achievements of the extended model and LJE simulation separately, and we finish with the outlook of this work.

9.1 About the *extended model*

We have formulated an algorithm that improves the pattern formation properties of the model originally proposed for ion beam sputtering by Cuerno, Makse, Tommassone, Harrington and Stanley (CMTHS) [Cuerno *et al.*, 1995]. This enhancement is necessary in order to have in the LJE simulation a reasonable representation of the etching front, as well as a relatively wide range of ripple wavelengths. This has been done by means of an improved estimation of the curvature at each site, which allows the erosion rule to be sensitive to smaller curvatures (large curvature radii), favoring the growing of ripples with longer wavelengths. Due to the averaging over the selected site and its nearest-neighbors, the estimated derivatives of the extended model provide erosion probabilities p_e less dependent on local fluctuations, favoring the emergence and growth of instabilities. Therefore, regular ripples with lengths between $30 \sim 50$ a units are obtained at earlier stages; whereas for similar times, the CMTHS model produces irregular ripples with a characteristic length of 10 a units. In general the ripple regime of the extended model occurs earlier and lasts longer which give opportunity to have a wider range of ripple lengths due to the coarsening phenomena.

In order to verify that the extended model inherits the properties of the CMTHS model, we compared the scaling properties of the interface width $W(t)$ for both models in their full versions (using the yield function $Y(\varphi) = 0.5 - 0.479\varphi^2 + 0.979\varphi^4$). For this comparison the parameter set of the extended model has been selected to be as close as possible to the original CMTHS model. Although not identical, both scalings present similar behavior, which shows that in spite of the modifications introduced in the extended model, both models are closely related.

Lauritsen et al. [Lauritsen *et al.*, 1996] derived a continuum Langevin equation for the CMTHS model. The resulting expression is a modified noisy Kuramoto-Sivashinsky (KS) equation. We have proceeded in a similar way with the extended model, and applying the method proposed by Vvendensky and others, we found that under the small gradients approximation, the extended model can be related to a gradient expansion whose leading terms also constitute a modified noisy KS equation. We have shown how this procedure can be also generalized to the LJE simulation, where in a reference frame moving with the laser, the etching is represented by a localized function (e.g. a Gaussian or an etching probability function). The resulting continuum equation is a modified noisy Kuramoto-Sivashinsky equation in a comoving coordinate frame (see [Mora *et al.*, 2005b]). In summary, the universal character of the patterns appearing in ion beam sputtering, water jet cutting, and LJE is also supported by the fact that their associated continuous stochastic differential equations are all related with the Kuramoto-Sivashinsky equation.

An important feature of the temporal evolution of the extended model, specially for the variant used in the LJE simulation ($Y(\varphi)=1$), is that once the ripple regime is established, the characteristic ripple wavelength grows with time due to ripple merging. This *coarsening* phenomenon is common to various natural and theoretical systems (see [Chakrabarti & Dasgupta, 2004; Politi & Misbah, 2004; Werner, 1999] and references therein). The increasing ripple wavelength is a deviation from systems described by the basic KS equation, where the selected ripple length scale remains constant (before the full non-linear regime that roughens the interface). Therefore, the coarsening effect can be attributed to high order terms of the modified noisy Kuramoto-Sivashinsky equation that corresponds to the extended model (see [Politi & Misbah, 2004; Raible *et al.*, 2000]).

9.2 About *LJE simulation*

Our basic assumption is that the erosion and surface diffusion rules of the extended model represent a physically relevant combination of processes to simulate pattern formation occurring in LJE. In the case of ion beam sputtering, the curvature dependency of the erosion rate has been derived from a microscopic analysis of the interaction of the incoming ion with the sample [Bradley & Harper, 1988; Sigmund, 1969]. Whereas in the LJE case, we have only presented a phenomenological justification. Of course this issue requires more investigation.

The LJE structuring process is driven mainly by the heat transport produced by the laser absorption. Our proposed *thermally thin layer* approximation is an oversimplification of the heat transport problem. Nevertheless, it provides a meaningful representation of the dependence of the temperature field on the laser power. Based on this temperature field, and taking into account that the etching only occurs above certain temperature threshold (due to the passivation layer), we have defined the *etching probability distribution*, which determines the spatial dependence of the rate of application of the model rules. The etching probability distribution provides a way to represent the fact that etching occurs in a region much broader than the laser spot. Up to our knowledge, the inclusion of a temperature field and the localized application of the model rules is an original contribution in the field of the solid-on-solid models for systems far from equilibrium. The lattice that represents the surface is kept fixed, while the etching probability distribution is scanned through the surface with a feed velocity v_f, creating an etching front that structures an 1+1 dimensional kerf. The dwell time depends on the width of the etching probability distribution and the feed velocity. The dwell time determines which of the different stages of the temporal evolution of the extended model are reproduced in the LJE simulation at the bottom of the kerf.

The main achievement of the LJE simulation is the existence (under specific parameter sets) of pattern formation regimes for varying feed velocity or varying laser power. More interestingly, there is a qualitative agreement with both analyzed experimental series. The essential mechanism behind this is the coarsening phenomenon observed in the extended model. Because the ripple length increases with the dwell time, it is expected that for increasing laser power (which means increasing of the diameter of the etching probability distribution) the obtained ripple length is larger. Correspondingly, when the feed velocity is increased, the dwell time decreases together with the ripple length.

Analyzing the different experimental series, we have made the simple observation that for an unstable etching front, the resulting ripple length is proportional to the average size of the etching front. The laser power determines the diameter of etching front, thus the resulting ripple length increases with power. We call this the *working hypothesis*, and it is consistent with the simulations: the diameter of the etching probability distribution and the length of the resulting ripples increase monotonically with the laser power.

9.3 Outlook

This work shows how a conceptually simple discrete model can be used in the investigation of the complex behavior that emerges in a non-linear system, such as the LJE experiment. Assuming that the observed pattern formation features are universal, the model uses various experimental and theoretical facts found in other contexts. Conversely, given that the model is not dependent of the microscopic details of the real etching, it could be adapted and applied to other techniques of localized structuring, like water jet cutting or laser cutting.

The flexibility of the kinetic Monte Carlo method allows to test the effect of different variables of the system. Additional physical mechanisms and refinements of the model can be introduced progressively and compared with further experiments. On the other hand, the formulation of the corresponding continuous Langevin equations can complement the analysis of the system, establishing a bridge between experiment and theory.

An open question remains: which is the internal mechanism responsible for ripple formation?. Oscillations in the etching reaction rates can be originated from two sources: (i) the fluctuations in the etchant concentration within the Nernst diffusive layer due to complex transport processes. Or, (ii) the non-linear character of the heat processes involved (absorption, conduction, and radiation). Further investigations should consider to model physical processes occurring not only in the interface but also within the etchant jet and the bulk material including all the couplings between them. Namely, it would be interesting to include other parameters and mechanisms of the experiment like the Nernst diffusive layer, the etchant concentration field, the dynamics of the jet flow, the exothermic heat produced by the etching reactions, etc.

In what follows, we outline some areas of future work:

- The next logical step is to perform further experiments in order to precisely characterize the ripples regimes and to establish relationships between the main parameters of the system. It is necessary to improve the knowledge of the microscopic processes of the experiment in order to validate the use of the erosion and surface diffusion rules. The working hypothesis about the dependence of the ripple length with the average size of the etching front should be verified with experiments where the dimensions of the etching front can be readily varied and measured.

- Concerning the LJE simulations, a systematic exploration of the parameter space can reveal the conditions for the onset of unstable etching fronts. For a more realistic approach, a "yield function" that represents the angle dependency of the absorption of polarized light must be included and accompanied with the estimation of the resulting variation of the etching probability distribution. Another feature to be fully characterized is the coarsening of the ripples, which is the basic mechanism behind the qualitative agreement between ripple regimes of the simulations and the experiments.

- All the basic assumptions of the model, like for example the thermally thin layer approximation, or the fact that all the physical processes are represented by erosion and surface diffusion rules should be examined with new experimental evidence.

- In the Appendix we show the preliminary results of a generalization of the extended model and the LJE simulation to 2+1 dimensions, which could provide a more realistic comparison with the experiments. There we propose an *ad hoc* estimation of the local curvature over a discrete field of 2D+1 dimensions that can reproduce the curvature-related instability of the 1+1D model. We show how ripple regimes can be generated using a moving Gaussian distribution. The etching probability function and all the other features of the LJE simulation should be generalized in a straightforward manner.

- Finally, we propose to enhance this kind of phenomenological discrete models for a broader set of localized structuring techniques. The formulation of model rules based on more realistic microscope mechanisms and material properties could provide a unified framework to simultaneously describe structuring in the macroscale, roughness in the microscale, and eventually emerging pattern formation in the intermediate scale.

Appendix A

Apppendix: Generalization to 2+1D

A logical step is the generalization the extended model and the LJE simulation to 2+1 dimensions. This will provide a more realistic comparison with the kerfs structured with the LJE technique. Here we propose an algorithm to define an average curvature for points of a discrete 2+1D lattice, and the generalization of the erosion and surface diffusion rules. Then as preliminary results, we show how uniform and rippled kerfs can be obtained when the power laser parameter is varied.

A.1   The surface in 2+1D

The material to be eroded is represented in 2+1D by a lattice composed of cells which are cuboids with a square base of side a and height b, and the surface is represented by the integer valued height $h_{i,j}$, where $i = 1 \ldots, L_x$ and $j = 1 \ldots, L_y$. The system size $L = L_x \times L_y$. Periodic boundary conditions apply for the height $h_{i,j}$. For each vertical column, all the sites below the surface are occupied with cells, whereas all the sites above are empty. As in the 1+1D case overhangs are not allowed. The temporal evolution of this virtual surface takes into account rules for representing the erosion and the surface diffusion processes in the 2+1D surface. The program defines an initially flat surface, selects a site and invokes with probability f the erosion rule and with probability $1 - f$ the diffusion rule.

A.2 The erosion rule

The erosion probability p_e for a cell at the site i is estimated as the product $p_e = p_\kappa Y_i$. The quantity p_κ corresponds to the probability of being eroded depending on the curvature of the surface at the site and accounts for the unstable erosion mechanism that exists in the physical system. The value of p_κ is defined larger for positive curvatures than for negative ones.

In general, the curvature at a point on an smooth surface does not have a unique value. The curvature of the arcs defined by sectioning the surface with a plane varies as the sectioning plane is rotated about the normal direction. The curvature achieves a minimum and a maximum (which are in perpendicular directions) known as the principal curvatures (see : [Weisstein, 2000],[Casey, 1996]). On the other hand, our discrete representation of the surface does not allow to perform an analytical estimation of the principal curvatures. Obviously the estimation of derivatives is constrained to the x or y directions.

The dependence of the absorption with the angle of incidence is

$$Y_{i,j} = Y(\varphi_{i,j}) = y_0 + y_1 \varphi_{i,j}^2 + y_2 \varphi_{i,j}^4 \tag{A.1}$$

Where the angle of incidence is defined as the angle $\varphi_{i,j}$ that forms the laser with the normal vector to the surface in each point ij. The nonlinearity introduced by this absorption function becomes relevant at late regimes when large slopes develop, then the ripples will be strongly distorted and the surface will have a rough morphology.

The algorithm for the estimation of derivatives, the angles and the curvatures is based on a finite central differences method around a selected site. The first derivatives or gradients of the surface at a point $[i, j]$ is:

$$\nabla_i = (h_{i-1,j} - h_{i+1,j})/2a, \;\; \text{and} \;\; \nabla_j = (h_{i,j-1} - h_{i,j+1})/2a \tag{A.2}$$

The normal vector to the surface is computed as:

$$\vec{\mathbf{N}}_{i,j} = [\nabla_i, \nabla_j, -1] \tag{A.3}$$

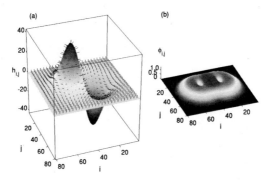

Figure A.1: . Estimation of the normal vector and incidence angles for a 2+1D surface of discrete values. (a) A surface of the type $h_{i,j} = i \exp(i^2 + i^2)$ is shown together with the estimation of the normal vectors with a method of finite differences. (b) Corresponding incidence angles, which are the angle that forms each normal vector with the vertical axis. The angle varies between 0 and $\pi/2$.

The normalized normal vector is

$$\hat{N}_{i,j} = (1 + (\nabla_i)^2 + (\nabla_j)^2)^{-\frac{1}{2}} [\nabla_i, \nabla_j, -1]. \tag{A.4}$$

Performing a dot product with the unitary vector in the vertical direction

$$\hat{N}_{i,j} \cdot [0, 0, 1] = (1 + (\nabla_i)^2 + (\nabla_j)^2)^{-\frac{1}{2}} = \cos(\varphi_{i,j}), \tag{A.5}$$

the angle of incidence can be computed

$$\varphi_{i,j} = \arccos\left[(1 + (\nabla_i)^2 + (\nabla_j)^2)^{-\frac{1}{2}} \right]. \tag{A.6}$$

In order to test the utility of these formulas the function $h_{i,j} = i \exp(i^2 + j^2)$ defined on a squared grid of points $[i, j]$ has been selected to represent a "valley" and a "peak". Figure A.1(a) shows the normal vector field resulting of applying the equation (A.4) for the function $h_{i,j}$. With these normal vectors and the equation (A.5) we estimated the corresponding incidence angles as it is shown in Figure A.1(b). Because the discreteness of the heights $h_{i,j}$, the values obtained with these formulas vary drastically from one site to the other. To attenuate this problem, the values for the angles and curvatures are computed not only for the site $[i, j]$ but also for its neighbors $[i - 1, j], [i + 1, j], [i, j - 1]$, and $[i, j + 1]$. This corresponds to a von Neumann neighborhood of range $r = 1$.

For a rough estimation of the curvature on a site $[i, j]$ we have to apply the nabla operator ∇ twice (see [Veseley, 2001]) :

$$\nabla^2 h_{i,j} \approx \frac{1}{a^2}(h_{i+1,j} + h_{i,j+1} + h_{i-1,j} + h_{i,j-1} - 4h_{i,j}) \tag{A.7}$$

This discrete Laplacian is almost constant when the a portion of the surface is close a paraboloid (which could eventually represent a valley or a peak of a ripple). Therefore the bare Laplacian can not be used as an estimation of the curvature. In order to differentiate clearly the curvature values for valleys and peaks, we propose an *ad hoc* generalization of the curvature formula used in the 1+1D case. For each lattice cell $[i, j]$ we have the first derivatives in each direction ∇_i and ∇_j. The gradient of the surface on the plane defined by ∇_i and ∇_j is :

$$\vec{\nabla}_{i,j} = \nabla_i \hat{\mathbf{x}} + \nabla_j \hat{\mathbf{y}} \tag{A.8}$$

In order to assign a unique value of curvature the lattice point $[i, j]$ we propose for the mean curvature:

$$\bar{\kappa}_{i,j} = \frac{\nabla_{i,j}^2}{(1 + |\vec{\nabla}_{i,j}|^2)^{\frac{3}{2}}} = \frac{\nabla_{i,j}^2}{(1 + (\nabla_i)^2 + (\nabla_j)^2)^{\frac{3}{2}}} \tag{A.9}$$

Note that when we impose $\nabla_j = 0$ we obtain the same formula than in the 1+1D case. Figure A.2(a) shows the computation of the discrete Laplacian $\nabla^2 h_{i,j}$ for the surface of Figure A.1 with the formula (A.7). Figure A.2(b) shows the estimation of the curvature for the same surface using the formula (A.9).

A.3 The diffusion rule

This rule is implemented as a generalization of the rule used in the extended model. The probability of a diffusive movement of a selected cell $[i, j]$ is evaluated selecting at random one of the von Neumann neighborhood (excluding the cell $[i, j]$) and computing the hopping probability (see [Siegert & Plischke, 1994])

$$w_{[i,j]}^{[i',j']} = \frac{1}{1 + \exp(\Delta \mathcal{H}_{[i,j] \to [i',j']}/(k_B T))}, \tag{A.10}$$

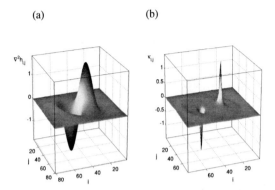

Figure A.2: (a) Laplacian of the surface of Figure A.1(a) estimate by a finite differences method (See equation A.7).(b) Estimate of the curvature of the same surface according to the equation A.9.

where $i' = \pm 1$ and $j' = 0$. When the selected diffusive movement is in the x-direction or $i' = 0$ and $j' = \pm 1$ when the selected diffusive movement is in the y-direction. Here $\Delta\mathcal{H}_{[i,j]\rightarrow[i',j']}$ is the energy difference between the final and initial states of the move, k_B is the Boltzmann constant and T is the temperature. This surface energy is defined through the Hamiltonian

$$\mathcal{H} = \frac{J}{b^2} \sum_{i,j=1}^{L_x,L_y} (h_{i,j} - h_{i+1,j})^2 + (h_{i,j} - h_{i,j+1})^2, \tag{A.11}$$

where J is a coupling constant through which the nearest neighbor sites interact and b is the height of the cells. Diffusion movements which produce final states with lower surface energy are then highly preferred.

A.4 A ripple regime

Using a two dimensional Gaussian distribution ($\sigma = 10a$) to represent the localized etching and varying the laser power, it is possible to find a ripple regime as in the 1+1D LJE simulation. Figure A.3 shows a kerf structured with relatively low power (10% of the maximum power). The bottom is shallow and presents a rough surface. The used laser

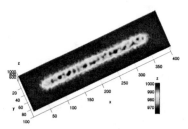

Figure A.3: Simulation of the structuring of a kerf in 2+1 dimensions with laser beam power 10%. The localized etching is represented by a Gaussian distribution with $\sigma = 10a$. The feed velocity is 1×10^{-6} per step. The erosion/diffusion invocation probability is $f = 0.015$ and for weighting the erosion probability the values $a = 4$, $p_{\kappa,min} = 0.15$, $c_{max} = 0.008$ have been used.

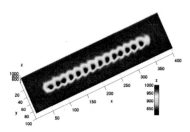

Figure A.4: Simulation of the structuring of a kerf in 2+1 dimensions with laser beam power 40%. The ripple length is approximately $20a$. The other model parameters are identical of Figure A.4.

feed velocity $v_f = 8 \times 10^{-7}$ a per step. The erosion/diffusion invocation probability is $f = 0.015$. For the computation of the derivatives $a = 4$ is used and the constant associated with the diffusion is $J/(k_BT) = 5$. For weighting the curvature dependent erosion probability p_κ the values $p_{\kappa,min} = 0.15$, $\kappa_{max} = 0.008$ have been used.

Figure A.4 shows ripples inside a kerf structured with all the same parameters as the previous case but with a laser power of 40%. Note that $f = 0.015$ means that the diffusion rule is applied 66.6 times more frequently than the erosion rule. This seems to be a strong requirment in the 2+1D case. The onset of the curvature dependent instability requieres that very smooth surface. Therefore the ripples will only occur after the time the diffusive rule had had enough time to smooth the surface. A slide of the kerf of Figure A.4 is shown

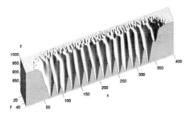

Figure A.5: Slide of the simulated kerf shown in Figure A.4.

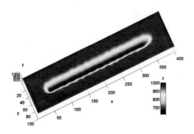

Figure A.6: A kerf with periodic ripples appear for laser beam power 100%. The other model parameters are identical of Figure A.3

in Figure A.5. The selected rendering of the surface allows to observe in the superior part a rough border product of the regions of the laser beam far from the center where the power is very low. The walls of the kerfs are ondulated and at the very bottom is possible to observe the maximum oscillation of the ripples. Figure A.6 shows what happens when the power is increased further up to 100% of the maximum value. Due to the effect of non linearities, the ripples are strongly damped and the kerf looks product of a approximately steady state etching front. However it is important to note that relatively small amplitude ripples remaing at the bottom of the kerf.

A.5 Discussion

The previous results show the possibility of generalize the extended and the LJE simulation to a 2+1 dimensions. Further work has to be done to incorporate a generalized etching probability distribution. Then, it is necessary to find a set of parameters to reproduce qualitatively the experiments of ripple regimes appearing when the feedvelocity or the laser power are varied. A further step could be the inclusion of the tridimensional dependency of the light absorption on the polarization and angle of incidence.

Bibliography

Anisimov, S. & Khokhlov, V.A. (1995). *Instabilities in Laser-Matter Interaction*. CRC Press, Boca Raton.

Barabási, A.L. & Stanley, H. (1995). *Fractal Concepts in Surface Growth*. Cambridge University Press, Cambrigde.

Bard, A. & Faulkner, L. (1980). *Electrochemical Methods. Fundamentals and Applications*. Wiley, New York.

Biehl, M. (2004). Lattice gas models and kinetic Monte Carlo simulations of epitaxial growth. *Invited lecture at the MFO Miniworkshop "Multiscale Modeling in Epitaxial Growth" (Oberwolfach 2004). Proceedings to be appear in "International Series in Numerical Mathematics" (Birkhaeuser)*, http://arxiv.org/abs/cond-mat/0406707.

Bradley, R.M. & Harper, J.M.E. (1988). Theory of ripple topography induced by ion bombardment. *Journal of Vacuum Science and Technology A: Vacuum, Surfaces, and Films*, **6**, 2390–2395.

Burton, W.K., Cabrera, N. & Frank, F.C. (1951). The growth of crystals and the equilibrium structure of their surfaces. *Philosophical Transactions of the Royal Society of London, Series A, Mathematical and Physical Sciences*, **243**, 299–358.

Carslaw, H.S. & Jaeger, J.C. (1959). *Conduction of Heat in Solids*. Oxford University Press, New York.

Carter, G. (2001). The physics and applications of ion beam erosion. *Journal of Physics D: Applied Physics*, **34**, R1–R22.

Casey, J. (1996). *Exploring Curvature*. Vieweg, Braunschweig.

Castro, M., Cuerno, R., Sánchez, A. & Domínguez-Adame, F. (1998). Anomalous scaling in a nonlocal growth model in the Kardar-Parisi-Zhang universality class. *Physical Review E*, **57**, R2491–R2494.

Castro, M., Cuerno, R., Vázquez, L. & Gago, R. (2005). Self-organized ordering of nanostructures produced by ion-beam sputtering. *Physical Review Letters*, **94**, 016102.

Chakrabarti, B. & Dasgupta, C. (2004). Mound formation and coarsening from a nonlinear instability in surface growth. *Physical Review E*, **69**, 011601.

Chua, A.L., Haselwandter, C.A., Baggio, C. & Vvedensky, D.D. (2005). Langevin equations for fluctuating surfaces. *Physical Review E*, 051103.

Collette, C. & Ausloos, M. (2004). Scaling analysis and evolution equation of the North Atlantic Oscillation index fluctuations. *International Journal of Modern Physics C*, **15**, 1353–1366.

Cross, M. & Hohenberg, P. (1993). Pattern formation outside of equilibrium. *Review of Modern Physics*, **65**, 851–1112.

Cuerno, R. & Barabási, A.L. (1995). Dynamic scaling of ion-sputtered surfaces. *Physical Review Letters*, **74**, 4746–4749.

Cuerno, R., Makse, H.A., Tomassone, S., Harrington, S.T. & Stanley, H.E. (1995). Stochastic model for surface erosion via ion sputerring : Dynamical evolution from ripple morphology to rough morphology. *Physical Review Letters*, **75**, 4464–4467.

Das Sarma, S. & Tamborenea, P. (1991). A new universality class for kinetic growth: One-dimensional molecular-beam epitaxy. *Physical Review Letters*, **66**, 325–328.

Datta, M. (1998a). Applications of electrochemical microfabrication: An introduction. *IBM Journal of Research and Development*, **42**, 563–566.

Datta, M. (1998b). Microfabrication by electrochemical metal removal. *IBM Journal of Research and Development*, **42**, 655–669.

Datta, M., Romankiw, L., Vigliotti, D. & von Gutfeld, R. (1987). Laser etching of metals in neutral salt solutions. *Applied Physics Letters*, **51**, 2040–2042.

Eden, M. (1961). *Two dimensional cellular growth*, vol. IV, chap. in Proceedings of the Fourth Berkeley Symposium on Mathematical Statistics and Probability. Contributions to Biology and Problems of Medicine. University of California Press, Berkeley.

Edwards, S. & Wilkinson, D. (1982). The surface statistics of a granular aggregate. *Proceedings of the Royal Society of London A*, **381**, 17–31.

Evans, M.R. (2000). Phase transitions in one-dimensional nonequilibrium systems. *Brazilian Journal of Physics*, **30**, 42–57.

Facsko, S., Bobek, T. & Kurz, H. (2002). Ion-induced formation of regular nanostructures on amorphous gasb surfaces. *Applied Physics Letters*, **80**, 130–132.

Facsko, S., Bobek, T., Stahl, A., Kurz, H. & Dekorsy, T. (2004). Dissipative continuum mode for self-organized pattern formation during ion-beam erosion. *Physical Review B*, **69**, 153412.

Family, F. & Vicsek, T. (1985). Scaling of the active zone in the eden process on percolation networks and the ballistic deposition model. *Journal of Physics A*, **18**, L75–L81.

Family, F. & Vicsek, T. (1991). *Dynamics of Fractal Surfaces*. World Scientific, Singapore.

Fernandes, M.G., Latanision, R.M. & Searson, P.C. (1993). Morphological aspects of anodic dissolution. *Physical Review B*, **47**, 11749–11755.

Finnie, I. & Kabil, Y.H. (1965). On the formation of surface ripples during erosion. *Wear*, **8**, 60–69.

Frank-Kamenetskii, D.A. (1969). *Diffusion and Heat Transfer in Chemical Kinetics*. Plenum Press, New York.

Friedrich, R., Peinke, J. & Renner, C. (2000a). How to quantify deterministic and random influences on the statistics of the foreign exchange market. *Physical Review Letters*, **84**, 5224–5227.

Friedrich, R., Radons, G., Ditzinger, T. & Henning, A. (2000b). Ripple formation through and interface instability from moving growth and erosion sources. *Physical Review Letters*, **85**, 4884–4887.

Georgescu, I. & Bestehorn, M. (2004). Pattern formation upon femtosecond laser ablation of transparent dielectrics, cond-mat/0411244.

Gerlach, C. (2002). *Räumliche und zeitliche Instabilitäten in einem technischen Prozess, Elektropolieren von Messing*. Ph.D. thesis, Bremen, Univ., http://elib.suub.uni-bremen.de/publications/ dissertations/E-Diss427_Gerlach_2002.pdf.

Gerlach, C., Visser, A. & Plath, P. (2004). Galvanostatic studies of an oxygen-evolving electrode. In G. Radons & R. Neugebauer, eds., *Nonlinear Dynamics of Production Systems*, 559–573, Wiley-VCH, Weinheim.

Glauber, R.J. (1963). Time-dependent statistics of the ising model. *Journal of Mathematical Physics*, **4**, 294–307.

Guo, W. & Johnson, D. (2003). Role of interfacial energy during pattern formation of electropolishing. *Physical Review B*, **67**, 075411.

Haase, M. & Widjajakusuma, J. (2003). Damage identification based on ridges and maxima lines of the wavelet transform. *International Journal of Engineering Science*, **41**, 1423–1443.

Haase, M., Widjajakusuma, J. & Bader, R. (2002). *Emergent Nature*, chap. Scaling laws and frequency decomposition from wavelet transform maxima lines and ridges, 365–374. World Scientific, Singapore.

Haase, M., Mora, A. & Lehle, B. (2004). Multifractal and stochastic analysis of electropolished surfaces. In M.M. Novak, ed., *Thinking in Patterns. Fractals and Related Phenomena in Nature*, 69–78, World Scientific, Singapore.

Habenicht, S., Lieb, K., Koch, J. & Wieck, A. (2002). Ripple propagation and velocity dispersion on ion-beam-eroded silicon surfaces. *Physical Review B*, **65**, 115327.

Haken, H. (1977). *Synergetics, An Introduction. Non Equilibrium Phase Transitions and Self Organization in Physics, Chemistry and Biology*. Springer-Verlag, Berlin.

Halping-Healy, T. & Zhang, Y. (1995). Kinetic rougnening phenomena,stochastic growth, directed polimers and all that. *Physics Reports*, **254**, 135–214.

Hartmann, A.K., Kree, R., Geyer, U. & Kölbel, M. (2002). Long-time effects in a simulation model of sputter erosion. *Physical Review B*, **65**, 193403.

Jafari, G.R., Fazeli, S., Ghasemi, F., Vaez Allaei, S., Reza Rahimi Tabar, M., Iraji zad, A. & Kavei, G. (2003). Stochastic analysis and regeneration of rough surfaces. *Physical Review Letters*, **91**, 226101.

Kardar, M., Parisi, G. & Zhang, Y.C. (1986). Dynamic scaling of growing interfaces. *Physical Review Letters*, **56**, 889–892.

Kim, J.M. & Kosterlitz, J.M. (1989). Growth in a restricted solid-on-solid model. *Physical review Letters*, **62**, 2289–2292.

Kratzer, P. & Scheffler, M. (2001). Surface knowledge: toward a predictive theory of materials. *Computing in Science and Engineering*, **3**, 16–25.

Krinsky, V.I., ed. (1984). *Self-Organization: Autowaves and Structures Far from Equilibrium*. Springer-Verlag, Berlin.

Krug, J. (1997). *Origins of scale invariance in growth processes*. Taylor & Francis, London.

Krug, J., Plischke, M. & Siegert, M. (1993). Surface diffusion currents and the universality classes of growth. *Physical Review Letters*, **70**, 3271–3274.

Kuramoto, Y. (1984). *Chemical oscillations, waves, and turbulence*, vol. VIII. Springer, Berlin.

Kuramoto, Y. & Tsuzuki, T. (1976). Persistent propagation of concentration waves in dissipative media far from thermal equilibrium. *Progress of Theoretical Physics*, **55**, 356–369.

Landau, D.P. & Binder, K. (2002). *A guide to Monte Carlo Simulations in Statistical Physics*. Cambridge University Press, Cambridge.

Lauritsen, K.B., Cuerno, R. & Makse, H.A. (1996). Noisy Kuramoto-Sivashinsky equation for an erosion model. *Physical Review E*, **54**, 3577–3580.

Lee, C., Takai, M., Yada, T., Kato, K. & Namba, S. (1990). Laser-induced trench etching of GaAs in aqueous KOH solution. *Applied Physics A*, **51**, 340–343.

Lind, P.G., Mora, A., Gallas, J.A.C. & Haase, M. (2005). Reducing stochasticity in the North Atlantic Oscillation index with coupled Langevin equations. *Physical Review E*, **72**, 056706.

Lu, Y. & Ye, K. (1996a). External-field-controlled laser wet etching of polycrystalline Al_2O_3TiC. *Applied Physics A*, **63**, 283–286.

Lu, Y., Takai, M., Nagatomo, S. & Namba, S. (1988). Wet-chemical etching of Mn-Zn ferrite by focused Ar+ -laser irradiation in H_3PO_4. *Applied Physics A*, **47**, 319–325.

Lu, Y.F. & Ye, K.D. (1996b). Laser-induced etching of polycrystalline Al_2O_3TiC in KOH aqueous solution. *Applied Physics A*, **62**, 42–49.

Lu, Y.F., Loh, T.E., Teo, B.S. & Low, T.S. (1994). Effect of polarization on laser-induced surface-temperature rise. *Applied Physics A*, 423–429.

Makeev, M.A. & Barabási, A.L. (2004a). Effect of surface morphology on the sputtering yields. I. ion sputerring from self-affine surfaces. *Nuclear Instruments and Methods in Physics Research B*, **222**, 316–334.

Makeev, M.A. & Barabási, A.L. (2004b). Effect of surface morphology on the sputtering yields. II. ion sputerring from rippled surfaces. *Nuclear Instruments and Methods in Physics Research B*, **222**, 335–354.

Makeev, M.A., Cuerno, R. & Barabási, A.L. (2002). Morphology of ion-sputtered surfaces. *Nuclear Instruments and Methods in Physics Research B*, **197**, 185–227.

Mallat, S. (2001). *Wavelet Tour of Signal Processing*. Academic Press, San Diego.

Marsili, M., Maritan, A., Toigo, F. & Banavar, J.R. (1996). Stochastic growth equations and reparametrization invariance. *Review of Modern Physics*, **68**, 963–983.

Meakin, P. (1998). *Fractals, Scaling and Growth Far from Equilibrium*. Cambridge University Press, Cambridge.

Meakin, P., Ramanlal, P., Sander, L.M. & Ball, R.C. (1986). Ballistic deposition on surfaces. *Physical Review A*, **34**, 5091–5103.

Mello, B., Chaves, A.C. & Oliveira, F. (2001). Discrete atomistic model to simulate etching of a crystalline solid. *Physical Review E*, **63**, 041113.

Metev, S., Stephen, A., Schwarz, J. & Wochnowski, C. (2003). Laser-induced chemical micro-treatment and synthesis of materials. *RIKEN Review: Focused on Laser Precision Microfabrication (LPM 2002)*, 47–52.

Metiu, H., Lu, Y. & Zhang, Z. (1992). Epitaxial growth and the art of computer simulations. *Science*, **255**, 1088–1092.

Momber, A.W. & Kovacevic, R. (1998). *Principles of abrasive water jet machining*. Springer, London.

Mora, A. & Haase, M. (2004a). Quantitative method for the characterization of complex surface structures. In *Nonlinear Dynamics of Electronic Systems (NDES 2004) as a special issue of the Journal: Nonlinear Dynamics*, Centro de geofísica de Evora, Evora (Portugal).

Mora, A. & Haase, M. (2004b). Wavelet analysis and scaling properties of electropolished surfaces. In F. Mallamace & H.E. Stanley, eds., *The Physics of Complex Systems (New Advances and Perspectives). Proceedings of the International School of Physics "Enrico Fermi". Course CLV*, 595–600, IOS Press, Amsterdam.

Mora, A., Gerlach, C., Rabbow, T., Plath, P. & Haase, M. (2004). Wavelet analysis of electropolished surfaces. In G. Radons & R. Neugebauer, eds., *Nonlinear Dynamics of Production Systems*, 575–592, Wiley-VCH, Weinheim.

Mora, A., Haase, M., Rabbow, T. & Plath, P. (2005a). Discrete model for laser driven etching and microstructuring of metallic surfaces. *Physical Review E*, **72**, 061604.

Mora, A., Rabbow, T., Lehle, B., Plath, P. & Haase, M. (2005b). A simple discrete stochastic model for laser-induced jet-chemical etching. In *Fractals in Engineering. New Trends in Theory and Applications*, Springer, London.

Muñoz-García, J., Castro, M. & Cuerno, R. (2005). Non-linear ripple dynamics on amorphous surfaces patterned by ion-beam sputtering. *http://arxiv.org/abs/cond-mat/0506469*.

Muraca, D., Braunstein, L. & R. C. Buceta, R.C. (2004). Universal behavior of the coefficients of the continuous equation in competitive growth models. *Physical Review E*, **69**.

Muzy, J.F., Bacry, E. & Arneodo, A. (1994). The multifractal formalism revisited with wavelets. *International Journal of Bifurcation and Chaos*, **4**, 245–302.

Newman, M.E.J. & Barkema, G.T. (1999). *Monte Carlo Methods in Statistical Physics*. Clarendon Press, Oxford.

Newport Corporation web page (2005). "Motion tutorial.". www.newport.com/file_store/Motion_Control/Tutorial/PDF_Files/motion_tutorial.pdf.

Nicolis, G. & Prigogine, I. (1977). *Self-Organization in Nonequilibrium Systems*. Wiley, New York.

Nowak, R. & Metev, S. (1996). Thermochemical laser etching of stainless steel and titanium in liquids. *Applied Physics A*, **63**, 133–138.

Ódor, G. (2004). Universality classes in nonequilibrium lattice systems. *Reviews of Modern Physics*, **76**, 663–724.

Politi, P. & Misbah, C. (2004). When does coarsening occur in the dynamics of one-dimensional fronts? *Physical Review Letters*, **92**, 090601.

Poupart, G. & Zumofen, G. (1992). Dissolution in (1+1) dimensions: a numerical study. *Journal of Physics A*, **25**, L1173–1179.

Press, W.H., Teukolsky, S.A., Vetterling, W.T. & Flannery, B.P. (1992). *Numerical Recipes in FORTRAN, the Art of Scientific Computing*. Cambridge University Press, Cambridge.

Prědota, M. & Kotrla, M. (1996). Stochastic equations for simple discrete models of epitaxial growth. *Physical Review E*, **54**, 3933–3942.

Rabbow, T. (2005). Private communication.

Rabbow, T., Mora, A., Haase, M. & Plath, P. (2005). Self-organized structure formation in organised microstructuring by laser-jet etching, (Accepted for publication in to nternational Journal of Bifurcation and Chaos in Applied Sciences and Engineering).

Radons, G., Ditzinger, T., Friedrich, R., Henning, A., Kouzmichev, A. & Westkämper, E. (2004). Nonlinear dynamics and control of ripple formation in abrasive water-jet cutting. In G. Radons & R. Neugebauer, eds., *Nonlinear Dynamics of Production Systems*, 391–410, Wiley-VCH, Weinheim.

Raible, M., Linz, S.J. & Hänggi, P. (2000). Amorphous thin film growth: Minimal deposition equation. *Physical Review E*, **62**, 1691–1705.

Ready, J.F. (1971). *Effects of High-Power Laser Radiation*. Academic Press, New York.

Renner, C., Peinke, J. & Friedrich, R. (2001). Experimental indications for markov properties of small-scale turbulence. *Journal of Fluid Mechanics*, **4**, 383–409.

Rost, M. & Krug, J. (1995). Anisotropic Kuramoto-Sivashinsky equation for surface growth and erosion. *Physical Review Letters*, **75**, 3894–3897.

Rusponi, S., Costantini, G., Boragno, C. & Valbusa, U. (1998). Scaling laws of the ripple morphology on Cu(110). *Physical Review Letters*, **81**, 4184–4187.

Sapoval, B., Santra, S. & Bardoux, P. (1998). Fractal interfaces in the self-stabilized etching of random systems. *Europhysics Letters*, **41**, 297–302.

Schmittmann, B. & Zia, R.K.P. (1995). *Statistical mechanics of driven diffusive systems*. Academic Press, London.

Schulz, W., Kostrykin, Vadim, N., Michel, J., Petring, D., Kreutz, E. & Poprawe, R. (1999). Dynamics of ripple formation and melt flow in laser beam cutting. *Journal of physics D: Applied physics*, **32**, 1219–1228.

Schwoebel, R. (1969). Step motion on crystal surface ii. *Journal of Applied Physics*, **40**, 614–618.

Shin, Y. & Jeong, S. (2003). Laser-assisted etching of titanium foil in phosphoric acid for direct fabrication of microstructures. *Journal of Laser Applications*, **15**, 240–245.

Siegert, M. & Plischke, M. (1994). Solid-on-solid models of molecular-beam epitaxy. *Physical Review E*, **50**, 917–931.

Sigmund, P. (1969). Theory of sputtering. I. sputtering yield of amorphous and polycrystalline targets. *Physical Review*, **184**, 383–416.

Sivashinsky, G.I. (1979). Hydrodynamic theory of flame propagation in an enclosed volume. *Acta Astronautica*, **6**, 631–645.

Smilauer, P., Wilby, M.R. & Vvedensky, D.D. (1993). Reentrant layer-by-layer growth: A numerical study. *Physical Review B*, **47**, 4119–4122.

Stephen, A. (2004). Private communication.

Stephen, A., Lilienkamp, T., Metev, S. & Sepold, G. (2002). Laser-assisted chemical micromachining of metals and alloys. *RIKEN Review : Focused on 2nd International Symposium on Laser Precision Microfabrication (LPM 2001)*, **43**, 56–62.

Stephen, A., Sepold, G., Metev, S. & Vollertsen, F. (2004). Laser-induced liquid-phase jet-chemical etching of metals. *Journal of Materials Processing Technology*, **149**, 536–540.

Takai, M., Lu, Y., Koizumi, T., Namba, S. & Nagatomo, S. (1988a). Thermochemical dry etching of single crystal ferrite by laser irradiation in CCl_4 gas atmosphere. *Applied Physics A*, **46**, 197–205.

Takai, M., Tsuchimoto, J., Tokuda, J., Nakai, H., Gamo, K. & Namba, S. (1988b). Laser-induced thermochemical maskless-etching of III-V compound semiconductors in chloride gas atmosphere. *Applied Physics A*, **45**, 305–312.

Tirumala Rao, B. & Nath, A.K. (2002). Melt flow characteristics in gas-assisted laser cutting. *Sadhana - Academy proceedings in engineering sciences*, **27**, 569–575.

Tok, E.S., Ong, S.W. & Kang, H.C. (2004). Dynamical scaling of sputter-roughened surfaces in 2 + 1 dimensions. *Physical Review E*, **70**, 011604.

Tutkun, M. & Mydlarski, L. (2004). Markovian properties of passive scalar increments in grid-generated turbulence. *New Journal of Physics*, **6**, 49.

Valbusa, U., Boragno, C. & Buatier de Mongeot, F. (2002). Nanostructuring surfaces by ion sputtering. *Journal of Physics : Condensed Matter*, **14**, 8153–8175.

Van Kampen, N.G. (1981). *Stochastic Processes in Physics and Chemistry*. North-Holland, Amsterdam.

Venkataramani, S.C. & Ott, E. (2001). Pattern selection in extended periodically forced systems: A continuum coupled map approach. *Physical Review E*, **63**, 046202.

Veseley, F.J. (2001). *Computational Physics: An Introduction*. Kluwer Academic/Plenum, New York.

Vold, M.J. (1959). A numerical approach to the problem of sediment volume. *Journal of Colloid Science*, **14**, 168–174.

von Gutfeld, R. & Sheppard, K. (1998). Electrochemical microfabrication by laser-enhanced photothermal processes. *IBM Journal of Research and Development*, **42**, 639–653.

von Gutfeld, R., Vigliotti, D. & Datta, M. (1988). Laser chemical etching of metals in sodium nitrate solution. *Journal of Applied Physics*, **64**, 5197–5200.

Vvedensky, D., Zangwil, A., Luse, C. & Wilby, M. (1993). Stochastic equations of motion for epitaxial growth. *Physical Review E*, **48**, 852–862.

Vvedensky, D.D. (2003a). Crossover and universality in the wolf-villain model. *Physical Review E*, **68**, 010601.

Vvedensky, D.D. (2003b). Edwards-wilkinson equation from lattice transition rules. *Physical Review E*, **67**, 025102.

Wächter, M., Riess, F., Schimmel, T., Wendt, U. & Peinke, J. (2004). Stochastic analysis of different rough surfaces. *The European Physical Journal B*, **41**, 259–277.

Waechter, M., Riess, F., Kantz, H. & Peinke, J. (2003). Stochastic analysis of surface roughness. *Europhysics Letters*, **64**, 579–585.

Walgraef, D. (1997). *Spatio-temporal pattern formation: with examples from physics, chemistry, and materials science*. Springer, New York.

Webb, C.E. & Jones, J.D.C., eds. (2004). *Handbook of Laser Technology and Applications*, vol. 3. Institute of Physics Publishing, Bristol.

Weeks, J.D. (1980). *Ordering in Strongly Fluctuating Condensed Matter Systems*, chap. The roughening transition, 293–317. Plenum Press, New York.

Weisstein, E.W. (2000). "curvature." from MathWorld - a wolfram web resource. Http://mathworld.wolfram.com/Curvature.html.

Werner, B.T. (1999). Complexity in natural landform patterns. *Science*, **284**, 102–104.

Wolf, D.E. & Villain, J. (1990). Growth with surface diffusion. *Europhysics Letters*, **13**, 389–394.

Yao, L. (2000). Laser machining processes. section 2.9: Reflection and absorption of laser beams. Http://www.columbia.edu/cu/mechanical/mrl/ntm/level2/ch02/html/l2c02s09.html.

Yewande, E.O., Hartmann, A.K. & Kree, R. (2005). Propagation of ripples in monte carlo models of sputter-induced surface morphology. *Phsyical Review B*, **71**, 195405.

Yuzhakov, V.V., Chang, H.C. & Miller, A.E. (1997). Pattern formation during electropolishing. *Physical Review B*, **56**, 12608–12624.

Yuzhakov, V.V., Takhistov, P.V., Miller, A.E. & Chang, H.C. (1999). Pattern selection during electropolishing due to double-layer effects. *Chaos*, **9**, 62–77.